ELECTRICAL ENGINEERING DEVELOPMENTS

HYBRID VEHICLES AND HYBRID ELECTRIC VEHICLES

NEW DEVELOPMENTS, ENERGY MANAGEMENT AND EMERGING TECHNOLOGIES

ELECTRICAL ENGINEERING DEVELOPMENTS

Additional books in this series can be found on Nova's website under the Series tab.

Additional e-books in this series can be found on Nova's website under the e-book tab.

ELECTRICAL ENGINEERING DEVELOPMENTS

HYBRID VEHICLES AND HYBRID ELECTRIC VEHICLES

NEW DEVELOPMENTS, ENERGY MANAGEMENT AND EMERGING TECHNOLOGIES

HILDA BRIDGES
EDITOR

nova publishers
New York

Library of Congress Cataloging-in-Publication Data

ISBN: 978-1-63482-157-5

Published by Nova Science Publishers, Inc. † New York

CONTENTS

PREFACE

With the ever-increasing worldwide demand for energy, and the looming crisis in petroleum supplies, energy storage is emerging as an important area of research. Due to ever increasing concerns on energy conservation and environmental protection, the hybrid vehicle (HV) is a widely accepted interim solution for evolving from the conventional internal combustion engine (ICE) vehicle to the clean electrified vehicle. This book discusses new developments, energy management and emerging technologies of hybrid vehicles and hybrid electric vehicles.

Chapter 1 – This chapter describes the state of the art in the field of Ultracapacitors (a.k.a. Super Capacitors), particularly as utilized at the present time in electric and hybrid vehicles. By comparison with batteries, ultracapacitors offer the advantages of very short charge/discharge time, virtually unlimited cycle life, zero maintenance requirements, and operability over a very wide range of temperatures. Ultracapacitors, however, still lag behind batteries in the aspect of energy density. Current research efforts to close that so-called "energy gap", which will allow ultracapacitors to be competitive with batteries, are described. The chapter also lists the key commercial and academic players in the area of ultracapacitor development, and describes trends and future expectations for the technology.

Chapter 2 – Hybrid vehicles are becoming increasingly available in the market, emphasizing the importance of a better understanding of its benefits in different driving conditions. Consumers have a distinct variety of hybrid designs available and this work intends to explore the differences between the two hybrid vehicle configurations (parallel/series and parallel configurations), based on a total of over 13 hours of 1 Hz real-world monitoring data. Five vehicles were monitored on-road and under real-world driving conditions, in Lisbon (Portugal). The vehicles were monitored with a Portable Emission

Measurement System to collect second-by-second information on engine parameters, tailpipe emissions and road topography. The data collected was analyzed using the Vehicle Specific Power (VSP) methodology to perform an energy and environmental characterization of the vehicles.

The parallel/series configurations present lower fuel consumption for lower VSP modes, while the parallel configurations are more efficient for higher VSP modes. While parallel/series configuration can only use the electric motor to move the vehicle under low power conditions (up to 11 to 12 W/kg depending on the vehicle) and turn the ICE off during a considerable amount of the time spent on braking, deceleration and idling, the parallel configuration only turns the ICE off at idling and only in a small part of the braking and deceleration time.

However, the electric motors are used to assist the ICE under higher power conditions (such as accelerations and hard starts). Therefore, these hybrid configurations present a trade-off, where the parallel/series configuration aims at reducing liquid fuel use mostly at low power conditions, while parallel configuration aims to reducing the liquid fuel use under high power conditions. Consequently, the energy and environmental performance of these vehicles is very dependent on the driving context.

Parallel/series hybrids present the lowest fuel consumption for the urban cycle, presenting, on average -30% of fuel consumption compared with average energy use of parallel configurations. Regarding the extra-urban driving cycle, the results are vehicle dependent and there is not a clear trend concerning which hybrid design presents the best fuel economy. Under highway conditions, parallel configuration uses the electric motor to support the ICE under acceleration, presenting the lowest fuel consumption, circa 11% lower than the full hybrid configuration. These conclusions can be transposed for CO2 emission and were also quantified for HC and NOx.

Summarizing, this work emphasizes not only the real-world impacts of the different hybrid configurations available, but also how effective they perform under typical drive-cycles, with different characteristics.

Chapter 3 – In this chapter, three emerging advanced permanent-magnet brushless machines are presented for hybrid vehicles. By introducing different types of hybrid vehicles, the power management for these vehicles is briefly introduced. Then, based on the aforementioned hybrid vehicle types, three emerging machines are presented for application in these vehicles, namely an outer-rotor permanent-magnet vernier motor for electric vehicle in-wheel motor drive, a dual-rotor dual-stator magnetic-geared PM machine for power splitting in series-parallel-type vehicles, and a hybrid stator-PM machine using as an integrated-starter-generator for complex-hybrid-type vehicles.

The design equations, operating principle and analytical results of these machines are quantitatively elaborated and discussed.

In: Hybrid Vehicles
Editor: Hilda Bridges

ISBN: 978-1-63482-157-5

Chapter 1

ULTRACAPACITORS FOR ELECTRIC VEHICLES: STATE OF THE ART AND TECHNOLOGICAL TRENDS

Ezzat G. Bakhoum, PhD
University of West Florida, Pensacola, Florida, US

ABSTRACT

This chapter describes the state of the art in the field of Ultracapacitors (a.k.a. Super Capacitors), particularly as utilized at the present time in electric and hybrid vehicles. By comparison with batteries, ultracapacitors offer the advantages of very short charge/discharge time, virtually unlimited cycle life, zero maintenance requirements, and operability over a very wide range of temperatures. Ultracapacitors, however, still lag behind batteries in the aspect of energy density. Current research efforts to close that so-called "energy gap", which will allow ultracapacitors to be competitive with batteries, are described. The chapter also lists the key commercial and academic players in the area of ultracapacitor development, and describes trends and future expectations for the technology.

Introduction and State of the Art

With the ever-increasing worldwide demand for energy, and the looming crisis in petroleum supplies, energy storage –particularly for transportation applications– is emerging as an important area of research. A relatively new component known as the Ultracapacitor (a.k.a. Super Capacitor) has gained much attention recently.

By comparison with batteries, ultracapacitors offer the advantages of very short charge/discharge time, virtually unlimited cycle life, zero maintenance requirements, and operability over a very wide range of temperatures.

Ultracapacitors, however, still lag behind batteries in the aspect of energy density. The energy storage capability of commercially available ultracapacitors is about an order of magnitude lower than lithium-ion batteries of the same dimensions, for example.

Ultracapacitor manufacturers are numerous. The largest manufacturers at the present time are: Maxwell Corporation, Eaton (formerly Cooper Bussmann Inc.), Illinois Capacitor Inc., Nesscap, and Panasonic. The following table lists a number of commercially available ultracapacitor products and their characteristics.

Table 2 lists the current energy storage requirements for various types of vehicle electric drive systems.

The ultracapacitor was introduced by Rightmire in 1966 [2]. During the past 15 years, much research on ways to improve the energy density in the Rightmire ultracapacitor have been published [3-8] and patented [9-13]. All the research activities prior to 2009, however, have focused primarily on improving the packaging of the Rightmire device. Innovative packaging techniques include the manufacturing of activated carbon electrodes with high volumetric efficiency [9, 10], and the replacement of conventional electrodes with carbon nano-tube electrodes, which offer higher surface areas [5-7].

Those efforts have been successful to a large degree, and ultracapacitors that offer capacitance densities in the range of 5 to 10 Farads/cm^3 are currently available commercially.

Packaging techniques, however, have reached their limit in terms of what the fundamental technology is capable of achieving. In 2009, the author of this article introduced a new technology for increasing the energy density of ultracapacitors by 1 to 2 orders of magnitude, thereby matching or exceeding the energy density of lithium-ion batteries. The author's research was published in the journals *IEEE Transactions on Ultrasonics, Ferroelectrics, and Frequency Control* [14] and the *Journal of Applied Physics* [15].

Table 1. A number of commercially available ultracapacitor products and their characteristics

Device	Rated Voltage	Capacitance (F)	Energy Density Wh/kg	Weight (kg)	Volume (lit)
Maxwell 	2.7	2885	4.2	.55	.42
Nesscap 	2.7	3640	4.2	.65	.51
Panasonic 	2.1	70	2.3	0.02	0.01
Illinois Capacitor 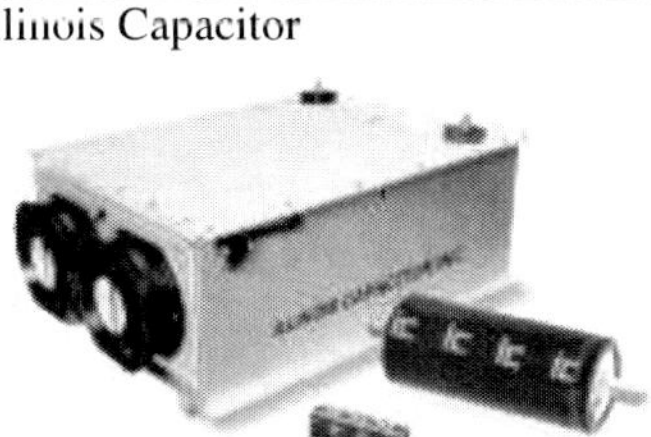	2.7	3800	5.7	0.675	0.47

Table 2. Energy storage requirements for various types of electric drive

Vehicle Drive System	System Voltage	Useable Energy Storage	Useable Depth of Discharge	Cycle Life (number of cycles)
Electric	300-400	15-30 kWh	Deep (70-80%)	2000-3000
Plug-in-hybrid	300-400	6-12 kWh battery 100-150 Wh Ultracapacitors	Deep (70-80%)	2500-3500
Micro-hybrid	45	30-50 Wh Ultracapacitors	Shallow (5-10%)	300k – 500k

From Burke [1].

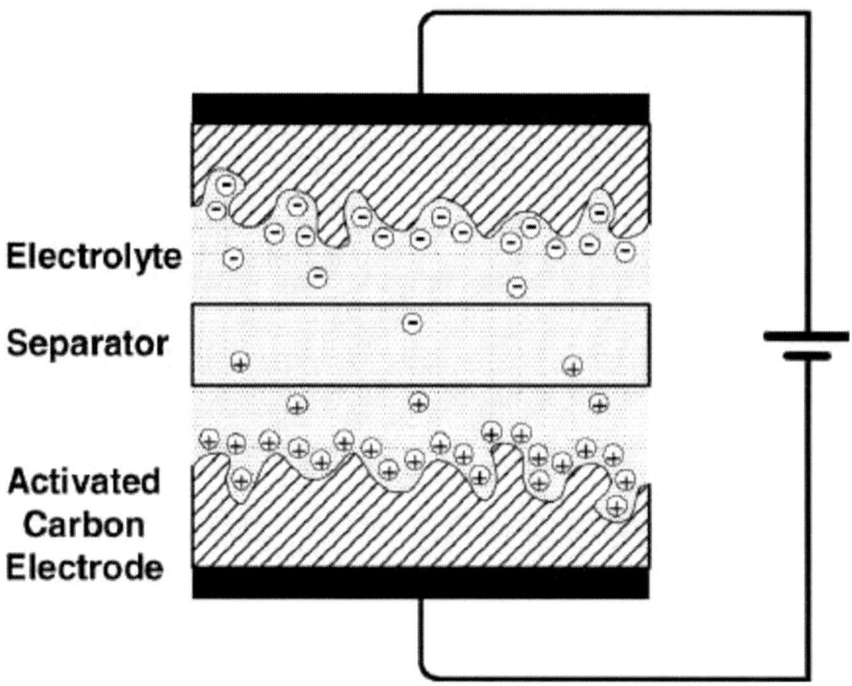

Figure 1. Cross-sectional view of an electrolytic double layer ultracapacitor.

During the years 2010-2014, numerous international research groups have published results [16-20] confirming the results published by the author in 2009.

Fundamentally, an ultracapacitor consists of two metal plates separated by an insulator, just like an ordinary capacitor. The separator, however, is porous and is soaked in an electrolyte. Since ions that form in the electrolyte can move freely through the separator, positive and negative ions move in opposite directions and cling to their respective electrodes (see Figure 1). The important feature in ultracapacitors is that the inner surface of each electrode is not a smooth surface but is rather padded with activated (porous) carbon. This results in a surface area that is about 100,000 times as large as the surface area of an ordinary capacitor. The immense surface area of an ultracapacitor, however, is not the only novel feature of the device. Since charges are now carried by ions that cling to the inner surfaces of the electrodes, the practical distance between the positive and the negative charges at each electrode is on the order of nanometers (the size of a molecule)[1]. The capacitance of a parallel-plate capacitor is given by the well-known equation [21]

$$C = \frac{\epsilon_0 \epsilon_r A}{d}$$

where ϵ_0 is the permittivity of free space, ϵ_r is the relative permittivity (or dielectric constant) of the dielectric present between the electrodes, A is the electrode area, and d is the distance between the electrodes. By maximizing A and minimizing d, therefore, ultracapacitors achieve extremely high values of capacitance (a capacitance on the order of several thousand farads is typical [1]). The technology published by the author in 2009 introduced for the first time a fundamental change to the structure of the ultracapacitor [14]. While the distance d between the positive and negative charges is minimized in a conventional ultracapacitor, the overall permittivity is only approximately that of free space (since $\epsilon_r \approx 1$ for an electrolyte thickness of a few nanometers). In the new technology introduced by the author (see Figure 2), a layer of a high dielectric-constant ceramic (B) is inserted as an interface between the electrolyte (C) and the surface of the activated carbon electrode (A).

For example, one particular ceramic that was used is $(Ba,Sr)TiO_3$, with a grain size of 50-100 nm and a relative permittivity ϵ_r of 10,000 or higher. While the insertion of such a layer reduces the basic capacitance per unit area 50-100 fold (due to the increased distance d), a gain in the overall permittivity

[1] The ultracapacitor has also earned the name "Electrolytic Double Layer Capacitor", or EDLC, due to the fact that two layers of charges exists near each electrode (see Figure 1).

of 10,000 or more is obtained. The result therefore is an increase of about 100-200 fold in the capacitance per unit area of the device.

As pointed out, ultracapacitors that are currently available commercially typically offer capacitance densities in the range of 5-10 Farads per cubic centimeter. Testing of new ultracapacitor samples that were assembled by the author according to the structure shown in Figure 2 has shown a capacitance in the range of 100-500 F/cm^3 (i.e., a gain of 10-50 fold over the commercial products, which is the minimum that the technology is capable of achieving. The theoretical limit will be approximately 100-200 fold as mentioned above).

As is well-known at present, the highest energy density offered by commercially available ultracapacitors (or EDLCs) is about 0.04 kWh/liter [1], or 0.15 kWh per volumetric gallon. A very basic calculation shows that the energy density in the new ultracapacitors introduced by the author exceeds 1 kWh/gallon on the average; which makes the new ultracapacitor technology directly competitive with lithium-ion batteries. Figure 3 shows a diagram illustrating the energy density vs. the power density for a number of technologies, including batteries (Li-ion, Ni-H, Lead-acid), fuel cells, the EDLC, and the new ultracapacitor introduced by the author. The successful manufacturing and commercialization of this new ultracapacitor structure will have important consequences for the transportation industry, due to the well-known advantages of ultracapacitors by comparison with batteries.

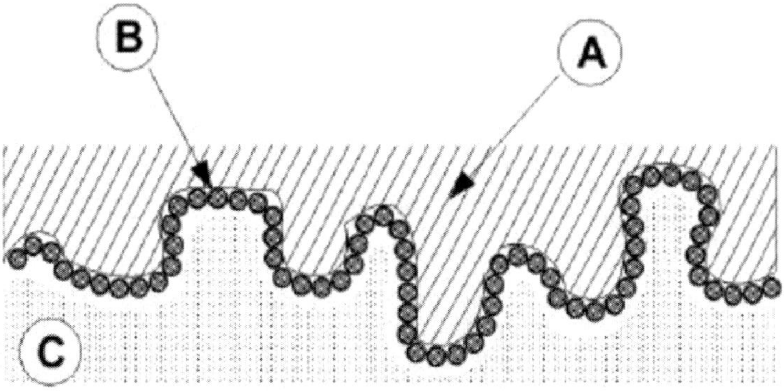

A: Activated Carbon. B: Nanometer-Sized Barium Titanade. C: Electrolyte.

Figure 2. Ultracapacitor structure introduced by the author (interface shown at one electrode only).

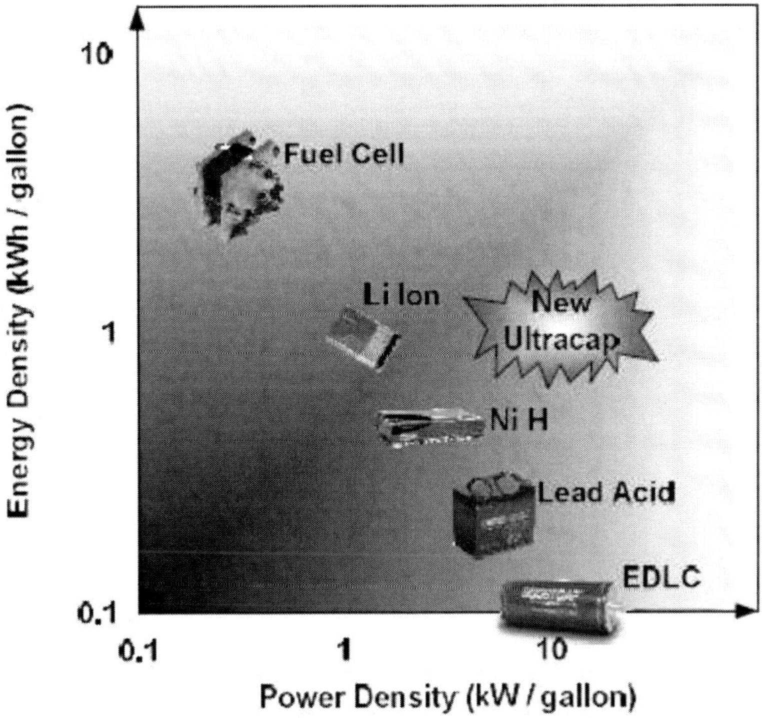

Figure 3. Energy density vs. Power density for various energy storage technologies.

CURRENT RESEARCH INVESTIGATIONS

The main results of the author's research were published in January of 2009 [14]. This was the first publication in which the concept of coating the activated carbon electrodes in the ultracapacitor with a high dielectric-constant material was shown. Subsequent to that publication, the following publications appeared in the literature:

[2010]:

- Luo et al. [16] replicated the technology shown in Figure 1 precisely in 2010 and reported the same capacitance values reported by the author.

- Chen et al. [17] demonstrated the same results by using a different coating material (MnO_2, which tends to have higher permittivity in the 50-100 nm range).
- Wang et al. [18] used a sophisticated technique in which oxide-covered Polyaniline was used as a coating material for the carbon electrodes, and reported a capacitance density of 950 F/cm^3 !

[2011]:

- Cheng et al. [19] once again repeated the same investigations by using MnO_2 as a coating material and reported capacitance densities in the range of 200 – 300 F/cm^3.

[2013]:

- The same results were demonstrated again in 2013 by He et al. [20].

For the ultimate success of the ultracapacitor structure shown in Figure 2 as a commercial energy storage product, it is critical to gain better understanding of the interface between the carbon electrode and the ceramic, as the chemical and physical properties of that interface may degrade over time. Theoretical and experimental investigations are now being conducted by the author and other groups in order to study, characterize, and eliminate such undesirable effects. A second goal of the ongoing investigations is to achieve optimization of the new technology, mainly through theoretical and experimental searching for the ultimate ceramic compound and the ultimate electrolyte that can be used in this application. Finally, the author and other research groups are currently conducting theoretical and experimental investigations of the process of scaling-up the technology. This effort is attempting to bridge the gap between the proof-of-concept phase and the subsequent large-scale development phase, which normally requires a mature technology. The following is a more detailed description of each of the research focus points and approaches listed above:

Ferroelectricity at the Nanometer Scale

The operation of the ultracapacitor structure shown in Figure 2 depends on the availability of a ceramic material with very high relative permittivity (ϵ_r).

It has been known for at least two decades that the permittivity of $BaTiO_3$ is a non-monotonic function of grain size [22].

For instance, Tsurumi et al. [23] measured the permittivity of pure $BaTiO_3$ as a function of the particle size and determined that the permittivity peaks at a value of about 5,000 for a particle size of approximately 140 nm (Figure 4). For the $(Ba,Sr)TiO_3$ ceramic, the value of the permittivity exceeds 10,000 for particle sizes in the range of 50-100 nm [24]. Other ceramics exhibit even higher values of permittivity. $PbMgNbO_3$ has a permittivity of 22,000 or higher for a particle size of 60-80 nm, as reported in [25] and [26]. More recently, giant and colossal permittivity has been observed in oxygen-deficient $BaTiO_3$ and $BaLaTiO_3$ ($\epsilon_r > 10^6$, at 300 °K and 100 Hz). Giant and colossal permittivity in those materials has been attributed to interfacial polarization, i.e., the transfer of charge carriers across the particle or grain. At the present time, ceramic materials with relative permittivities in the range of 10^4 to 10^6 (with sizes in the range of 50-100 nm) are available commercially from specialized chemical suppliers.

Deposition of Nanometer-Sized Ceramic Particles on the Inner Surfaces of Activated Carbon or Carbon Nanotubes

The implementation of the concept shown in Figure 2 depends on the successful deposition of very small ceramic particles on the inner surfaces of the pores of an activated carbon electrode.

The problem of the deposition of particles of a ceramic material on the inner surfaces of porous carbon was investigated in the early 1990s by Haber et al. [27-31]. Haber et al. used the electrophoretic deposition technique [30, 31] and experimentally demonstrated the feasibility of deep penetration and deposition of colloidal particles of SiO_2 on the inner surfaces of porous graphite. Although the technique demonstrated by Haber et al. was not meant to be used in capacitor applications, it was adopted and used by the author without any difficulty [14, 15]. Under the right conditions, deep penetration of nanometer-sized ceramic particles inside porous carbon and the deposition of such particles on its inner surfaces can be successfully accomplished. Figure 5 (a) is a scanning-electron microscope (SEM) figure that shows the deposition of particles of $(Ba,Sr)TiO_3$ (size 50-100 nm) deep inside the pores of an activated carbon electrode. The size of the pores in the electrode is in the range of 0.5 μm to 15 μm in the present application.

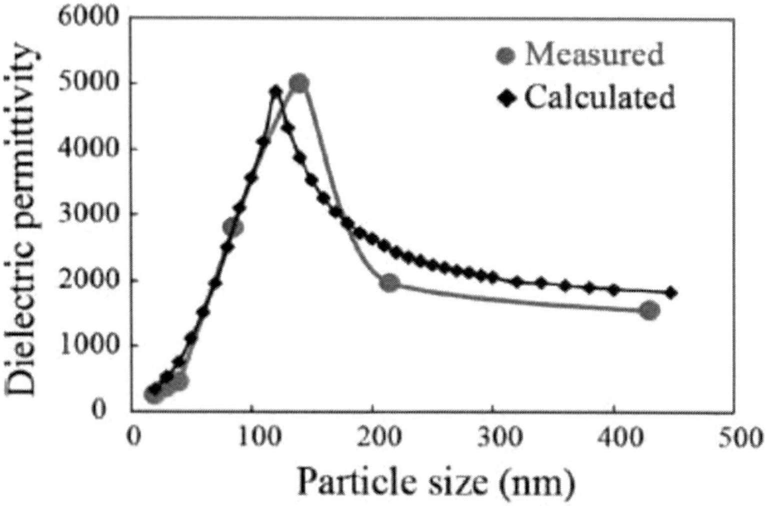

Figure 4. Dielectric constant v. particle size.

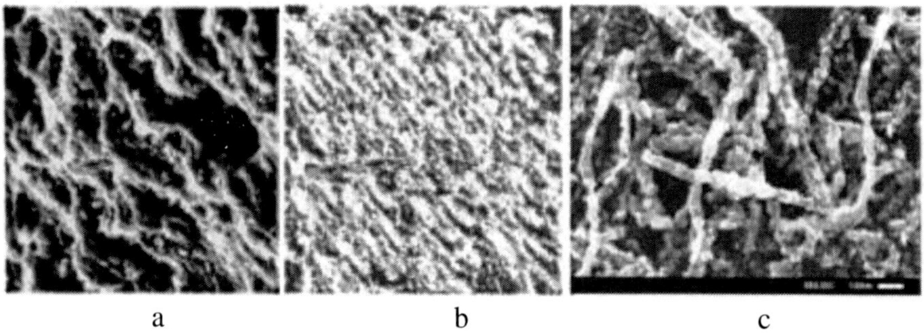

Figure 5. (a) Colloidal $(Ba,Sr)TiO_3$ deposited on the inner surfaces of a porous carbon electrode (view of a cross-section). (b) Deposition of the same ceramic under ideal electrophoretic deposition conditions. (c) Deposition of the same ceramic on the surfaces of carbon nanotubes of a diameter of 50 nm.

In Figure 5(a), the conditions for electrophoretic deposition were non-ideal (see Reference [14]), and the surface area covered was only 56%. Figure 5(b) shows the result that was obtained under ideal conditions (low electric field intensity and long exposure time, as demonstrated in [14]). The surface area covered in Figure 4(b) was 83%.

Typically, under ideal conditions for electrophoretic deposition, surface area coverage is in the range of 80% to 90%. This is the area of the electrode that exhibits substantial increase in capacitance.

The remaining uncovered area simply results in a circuit model in which a low-capacitance device appears in parallel with a very high-capacitance device. Figure 5(c) shows the same ceramic particles applied as a coating on the inner surfaces of an electrode made of carbon nanotubes (CNTs). The diameter of the CNTs is approximately 50 nm, and individual particles of the ceramic can be seen in the figure. As demonstrated in [14] and [15], different cross-sectionalSEM and TEM micrographs have showed that – by maintaining the right conditions for electrophoretic deposition – the coatings are quite uniform and the thickness of the deposited ceramic layer can be controlled with a high degree of precision. The electrophoretic deposition process is typically followed by sintering in order to obtain strong adhesion of the coating to the carbon surface. Electrophoretic deposition can be scaled up for mass manufacturing without difficulty [30], and other applications in the industry that use the process for the coating of various materials are well known.

The Composition of the Electrolyte and Its Interaction with the Ceramic Layer

The electrolytes that are commonly used in ultracapacitors at present usually consist of a solvent, such as acetonitrile or propylene carbonate, in which an ionic salt is dissolved. Acetonitrile is generally preferred due to its high dielectric strength and high ionic mobility. The most common ionic salt in use is tetraethylammonium-tetrafluoroborate (TEA/BF_4) [32-39]. In ordinary ultracapacitors, the anions and cations of the ionic salt (e.g., TEA^+ and BF^-) become solvated by the molecules of the solvent and are surrounded by such molecules [34-36]. This results in the anions and cations reaching a distance of approximately 1 nm (the size of a molecule) from the surface of the carbon electrode and maintaining that distance (and hence the very high capacitance of the EDLC per unit area, by comparison with traditional capacitors). In the new ultracapacitor introduced by the author, the surface of the activated carbon is covered with a layer of a highly polar ceramic, as previously indicated. Studies of the interaction of solvated ionic salts with polar ceramics exist in the chemistry and the electrochemistry literature [40-43]. According to those studies, the solvated ionic salts were found to form strong bonds with the particles of the ceramic. In the new ultracapacitor design shown in Figure 2, the solvated anions and cations bond strongly with the highly polar $(Ba,Sr)TiO_3$ ceramic, which effectively results in the ceramic layer acting as the new (and only) dielectric in the improved capacitor.

It is to be pointed out that the ionic salt used in the new ultracapacitor design has a concentration that is much higher than the concentrations typically used in ordinary ultracapacitors. For instance, in ordinary ultracapacitors, TEA/BF_4 is typically dissolved in the solvent at concentrations ranging from 4 M/gallon to 8 M/gallon.

In the new ultracapacitor, the capacitance per volumetric gallon has a value of about 10^6 F on the average. By using the well-known equation $Q = CV$ and Avogadro's constant, and given an operating voltage of 2.7 Volts, it is easily calculated that approximately 28 M of the ionic salt will be required in a volumetric gallon. At such high concentrations, a salt such as TEA/BF_4 is nearly insoluble and therefore nearly impossible to use.

Another ionic salt, specifically, lithium borohydride ($LiBH_4$), is much more efficient at such high concentrations because of its low molar weight (21.78 g). $LiBH_4$ is in fact the salt that was used in the new ultracapacitor samples that demonstrated very high capacitance per unit volume (100-500 F/cm^3).

Capacitance, Cyclic Voltammetry, Discharge Curve, and ESR

Figure 6(a) shows a plot of the measured capacitance of the new ultracapacitor samples, as reported in the author's publication in 2009 [14]. The capacitance is shown as a function of the exposure time and the electric field intensity used during the electrophoretic deposition process. As the graphs show, a lower electric field and longer exposure time leads to substantially better surface coverage and hence higher capacitance (the two graphs correspond to figures 5(a) and 5(b) above). Figure 6(b) shows cyclic voltammetry curves (the electrolyte used was acetonitrile).

The operating voltage of the new ultracapacitor is 2.7 V, just like ordinary ultracapacitors, and hence the cyclic voltammetry curves are very similar to those of existing ultracapacitors. Figure 6(c) shows the measured open-circuit voltage across the terminals of a sample over a period of 72 h (discharge curve). The internal leakage current was determined to be substantially lower than leakage currents in existing ultracapacitors.

However, a closed-circuit discharge test showed that the Equivalent Series Resistance (ESR) is about 0.5 mΩ on the average, which is comparable to existing ultracapacitors.

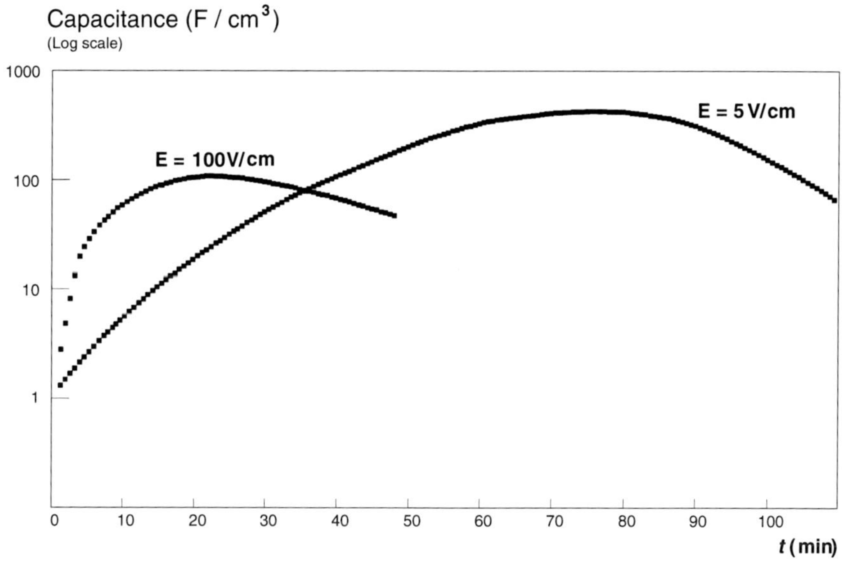

Figure 6 (a) Measured capacitance density (Farads per cubic centimeter), as a function of the applied electric field intensity and the exposure time during the electrophoretic deposition process.

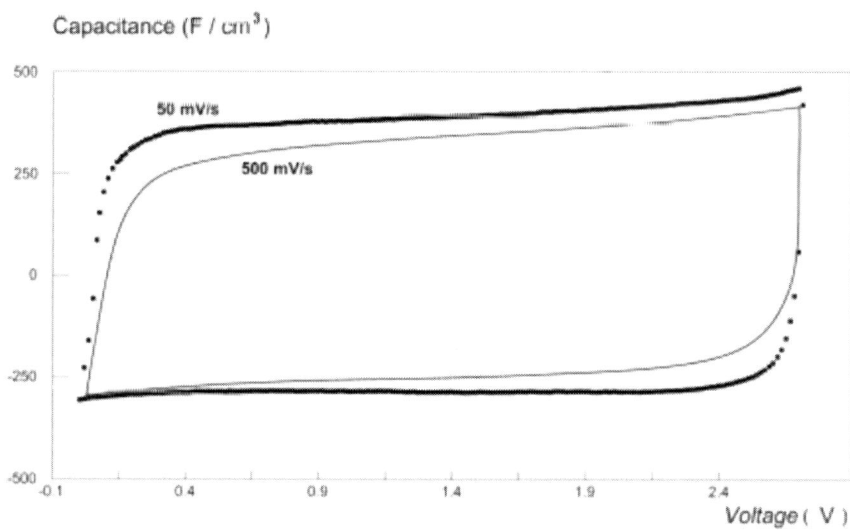

Figure 6 (b) Cyclic voltammetry at scan rates of 50 mV/s and 500 mV/s.

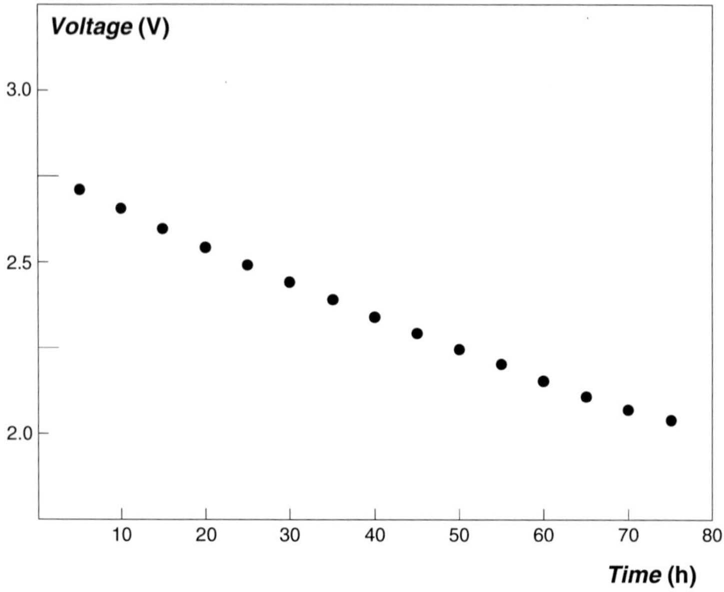

Figure 6 (c) Open-circuit voltage of the ultracapacitor, measured over a period of 72 hours.

Cost Analysis

At the present time, the electrical energy storage device that offers the highest energy density is the lithium-ion battery. The cost of lithium-ion batteries is $300 per kWh, according to published industry figures [44].

For ultracapacitors that are available commercially, the cost per Farad is approximately $0.005 [45-50] (and this cost is expected to fall significantly in the near future). Given that the new ultracapacitor introduced by the author packs more than 10 times the capacitance of ordinary ultracapacitors in the same volume, a very basic calculation shows that an ultracapacitor with a 1 kWh energy storage capability should cost about $494.

This is obviously higher than the cost of a comparable lithium-ion battery; however, with the efficiencies of mass-manufacturing the cost is expected to drop significantly as pointed out.

RESEARCH AND DEVELOPMENT PLANS FOR THE NEXT DECADE

1. Investigation of the Physical and Chemical Properties of the Carbon Surface/Ceramic Interface, the Stability of That Interface, and the Possible Enhancement of the Interface by Using Thin Film Structures

In practical applications, the new ultracapacitor introduced by the author must withstand thousands of charge/discharge cycles under widely varying temperatures. Ordinary ultracapacitors, as is well known, have succeeded in this regard. The reliability of the proposed new ultracapacitor, however, will mainly depend on the durability of the ceramic-coated electrodes.

One of the well-known challenges in capacitive structures using high-K dielectrics is the possible degradation of the dielectric/conductor interface over time. To ensure the durability of the electrodes described above under real-life conditions, it is necessary to investigate the stability of the dielectric/carbon surface interface, and the possibility of enhancing that interface by using thin-film structures [51].

As described further below, extensive experimental testing of the new ultracapacitors will first be conducted. The particular concern in the present application is the possible oxidation of the carbon surface over time, which will lead to a degradation of the dielectric/carbon surface interface as indicated. Should testing reveal that such an effect is present, the author and his research team will investigate the insertion of barrier layers between the high-K dielectric and the surface of the carbon electrode. In previous published studies [52, 53], carbon nitride (CN_x) has been successfully used as a barrier layer due to its high stability and its resistance to corrosion. CN_x is a conductor, with a resistivity (0.006 - 0.04 Ω cm) that is actually lower than the resistivity of the activated carbon material (0.4 - 2 Ω cm) [54]. Hence, the addition of a CN_x layer as a barrier between the high-K ceramic and the activated carbon surface not only protects the surface and ensures its stability, but further actually enhances the electrical properties of the electrode. Other conducting barrier materials, such as Zr-Ge-N and ZrN [55], W-B-N films [56], and Ir/TaN bilayers [57] will also be investigated in order to identify viable barrier materials for the high-K dielectric/carbon interface structure.

The tool that the author plans to use in the investigations will be a Pulsed Laser Deposition (PLD) system (the author currently has access to a PLD system through the Major Analytical and Instrumentation Center (MAIC) at the University of Florida – see the Facilities and Equipment section, and the support letter attached to this proposal). PLD will be used for the deposition of both the barrier layer and the high-K ceramic. PLD provides atomic-level control of film growth and is particularly effective in the growth of perovskites, such as $BaTiO_3$. In the PLD apparatus that the author plans to use a focused laser pulse is directed onto a target of material in a vacuum chamber. The laser pulse locally heats and vaporizes the target surface, producing an ejected plasma or plume of atoms, ions, and molecules. The plume of material is deposited onto an adjacent substrate to produce a crystalline film. This technique possesses several favorable characteristics for the growth of multi-component materials, such as stoichiometric transfer of target material to the substrate and atomic-level control of the deposition rate. Ultimately, the CN_x layer – if required – will have to be deposited on the surface of the activated carbon or carbon nanotube electrodes during mass manufacturing.

For that purpose, Atomic Layer Deposition (ALD) has proved to be very suitable for the deposition of very thin films of nitride layers on a large scale [58], since the PLD technique is not cost effective in mass manufacturing. Furthermore, the author expects that electrophoretic deposition (EPD) will be ultimately the technique of choice for depositing the high-K ceramic in a mass manufacturing environment.

To test the ultracapacitor samples produced throughout the research, the author intends to acquire a programmable variable temperature chamber, which will be used for testing the ultracapacitors after repeated charge/discharge and temperature cycles[2], and intends to build circuitry for simulating real-life charge/discharge conditions for those ultracapacitors (see budget section). The author currently has access to Scanning Electron Microscopy and X-ray Diffraction equipment through the MAIC center at the University of Florida (see the Facilities and Equipment section and the support letter attached to this proposal).

The effect of the degradation of the dielectric/carbon surface interface, if present, will be readily observable with the available equipment. While the effect of incorporating C into the $(Ba,Sr)TiO_3$ lattice is not known at present, it is possible that C may substitute for Ti^{4+}.

---

[2] Temperature cycling is a well-known technique for simulating aging. A tyauthorcal battery or ultracapacitor that is exposed to environmental conditions may experience as much as 700 temperature cycles per year. Therefore, if testing shows that the proposed new ultracapacitor can withstand 7000 temperature cycles, then the expected lifetime of the device is approximately 10 years.

It is also possible that C may form CO or CO_2 which could react with $BaTiO_3$ to form a carbonate (e.g., $BaCO_3$) and a Ba-deficient oxide (e.g., $BaTi_2O_5$, TiO_2, etc.) [59]. Such phases will be apparent from X-ray diffraction examination of disassembled ultracapacitors. Of course, should those reactions prove to be present, a conductive barrier layer (e.g., CN_x layer) will be necessary as indicated above.

Another concern when high-K ceramics are used in capacitors is the possible variation of the dielectric constant (and hence capacitance) with the variation in temperature. Tsurumi et al. [23] recently measured such variation in capacitance for $BaTiO_3$ particles of a size of approximately 140 nm. The result is shown in Figure 7.

Very similar results for $(Ba,Sr)TiO_3$ thin films were published by Cheng et al. [60] and by Cole et al. [61]. Clearly, both $BaTiO_3$ and $(Ba,Sr)TiO_3$ are X7R dielectrics [62], with a maximum variation in capacitance of ± 15% over a temperature range of -55°C to +125°C. This is quite acceptable for the present application. The objective of the planned testing, therefore, will be rather to determine the stability of the dielectric/carbon surface interface as indicated.

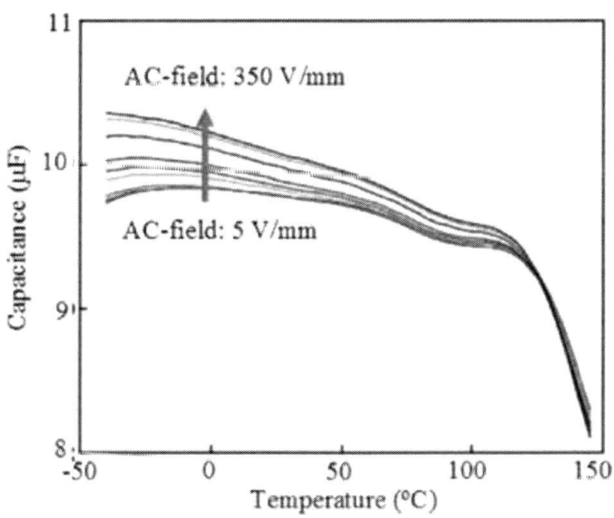

Figure 7. Variation of a nominal capacitance of 10 μF with temperature.

Figure 8. Electrophoretic deposition equipment used by the author.

2. Scaling up and Optimization of the Technology: Bridging the Gap between the Prototype Phase and the Manufacturing/ Large-Scale Development Phase

This section of the planned research effort will consist of 4 tasks:

a) Numerical computations: The main difference between the proposed new ultracapacitor and traditional ultracapacitors is the electrode structure (see Figure 2). To manufacture such electrodes in a cost-effective manner, EPD must be performed on a large (industrial) scale for the deposition of the ceramic layer on very large areas of activated carbon/carbon nanotube sheets in a reasonable amount of time. Numerical computations will be performed in order to determine the necessary conditions for such large-scale EPD coating of electrode surfaces. More specifically, computations based on the fundamental law of electro-osmotic flow [15] will be carried out to determine the conditions that will ensure maximum particle penetration inside the porous electrodes in the shortest amount of time. Such computations are not trivial, due to the various factors involved in the process (hydrodynamic forces, continuity equation, variable electrical permeability, etc. – see Reference [15]). Essentially, the computational problem will be a boundary-value problem that must be solved in order to determine the optimal arrangement/apparatus for large-scale, time effective coating of activated carbon/carbon nanotube surfaces. Such computational work will be performed on a high-end workstation by using a tool such as C++ or Matlab. Figure 8 above shows the EPD equipment that is currently used by the author for small-scale electrode surface coating. It must be pointed out that EPD has been successfully performed on an industrial scale in numerous other applications in the recent past. Wires of several kilometers of length have been successfully coated with oxide ceramics by using EPD [63]. Thin films of ZrO_2 have been deposited inside porous substrates [64]. Automobile bodies, appliances, power tools, and even superconducting materials have all been coated with EPD [65-67].

b) Experimentation with Atomic Layer Deposition (ALD): The author currently has access to ALD equipment through the MAIC at the University of Florida. This section of the investigation will focus on the coating of large surfaces of carbon/carbon nanotube sheets with conductive barrier layers, such as CN_x. This effort can be characterized as a low-risk effort, and will be performed mainly as a learning experience for the author and the research team (graduate and undergraduate students).

c) Experimental search for the optimal ceramic material: Substantially better ultracapacitors can be obtained if stable materials with higher dielectric constants are used. This section of the proposed research involves synthesizing high-permittivity ceramics and investigating the relationship between the dielectric constant (ϵ_r) and the macroscopic properties of the ceramic layer. In particular, the effect of material composition and grain size on ϵ_r will be rigorously determined. Several different materials will be investigated including $(Ba,Sr)TiO_3$, $PbMg_{1/3}Nb_{2/3}O_3$-$PbTiO_3$, MnO_2, in addition to newer materials that exhibit colossal permittivity [68-70]. Some of the powders will be obtained commercially and others will be synthesized. For example, BST will be synthesized by means of an oxalate precipitation route by using aqueous $BaCl_2$, $SrCl_2$, and $TiCl_4$ solutions in oxalic acid, followed by low temperature calcinations. This wet chemistry route has been shown to produce particles of approximately 20-50 nm in diameter [70]. The particle size can be increased by increasing the calcination temperature and hold times after the oxalate route. Several different particle sizes of $(Ba,Sr)TiO_3$ powders will be available from the different calcination temperatures and also those synthesized using standard solid state calcination of $BaCO_3$, $SrCO_3$, and TiO_2 powders. Several different ultracapacitor prototypes will be assembled using the different particle size distributions and material compositions for correlation to permittivity measurements. Dense, nanocrystalline ceramics will be prepared by Spark Plasma Sintering (SPS). The author has access to an SPS system through the MAIC center at the University of Florida. The density, microstructure, and dielectric permittivity of samples prepared via this method will be measured and studied with electrochemical impedance spectroscopy (EIS).

d) Theoretical an experimental search for an optimal electrolyte: Improvement of the electrolyte can also lead to potential improvement in the performance of the proposed new ultracapacitor. The author plans to investigate the possible use of ionic liquids in lieu of dissolved ionic salts in the new ultracapacitor design. Ionic liquids, such as BMIM/(BF_4, PF6, $N(CN)_2$, etc.) have attracted lots of attention recently because they offer higher electrochemical voltage windows [71-74]. According to the equation $E = ½ CV^2$, a doubling of the operating voltage of the ultracapacitor from 3 V to 6 V, for example, will result in the quadrupling of the stored energy. Ionic liquids, on the other hand, suffer from inherently poor disassociation of anions and cations [71], which usually results in lower capacitance per unit volume and a higher equivalent series resistance (ESR) of the ultracapacitor. However, for the new ultracapacitor structure described here, it is believed that the performance of ionic liquids may be substantially better, in view of the electrochemical studies [40-43] concerning the interfacial reactions of such species with polar ceramics. Essentially, a highly polar ceramic such as $(Ba,Sr)TiO_3$ is expected to increase the disassociation rate of anions and cations, which will lead to lower ESR and a higher capacitance density. Such effects will be investigated experimentally. A bigger plan, however, is to investigate the mentioned interfacial reactions theoretically by using the very powerful Density Functional Theory (DFT) [75-85]. The use of DFT will also allow the interfacial reactions of a wide array of ionic salts with polar ceramics to be investigated computationally. Pseudopotentials for the Local Density Approximation (LDA), which exist in the literature [86-88], are quite adequate for simulating interfacial reactions at the molecular or particle level.

e) The results of this study will lead to vastly better understanding of the behavior of ionic liquids and ionic salts near polar ceramics, and hence will help pinpoint the electrolyte that will optimize the energy density and the ESR in the new ultracapacitors. It is to be pointed out that DFT has been successfully used in the recent past for the study of interfacial reactions in other applications [89-91].

REFERENCES

[1] A. Burke, "Ultracapacitors: why, how, and where is the technology", *J. of Power Sources*, v. 91, 2000, pp. 37-50. (Updated in 2010: A. Burke, *Int. J. of Energy Research*, v. 34, no. 2, 2010, pp. 133-151).

[2] R. A. Rightmire, "Electrical Energy Storage Apparatus", *US Patent No. 3,288,641,* Nov. 29, 1966.

[3] B. E. Conway, "*Electrochemical Supercapacitors*", (Kluwer Academic Publishers, New York, NY, 1999).

[4] M. Wixom, I. Song, C. Peiter, J. White, and L. Thompson, "A survey of early transition metal nitride ultracapacitor electrodes", *Proceedings of the Electrochemical society*, v. 15, 1999, pp. 627-638.

[5] I. H. Kim, J. H. Kim and K. B. Kim, "Electrochemical Characterization of Electrochemically Prepared Ruthenium Oxide/Carbon Nanotube Electrode for Supercapacitor Application", *Electrochemical and Solid-State Letters,* v. 8, no. 7, 2005, pp. A369-A372.

[6] D. Chunsheng and N. Pan, "High power density supercapacitor electrodes of carbon nanotube films by electrophoretic deposition", *Nanotechnology*, v. 17, no. 21, 2006, pp. 5314-5318.

[7] S. Chakraborty et al., "Surface Area Measurement of Functionalized Single-Walled Carbon Nanotubes", *J. Phys. Chem. B*, 110, 49, 2006, pp. 24812-24815.

[8] G. Lota, E. Frackowiak, J. Mittal, and M. Monthioux, "High performance supercapacitor from chromium oxide-nanotubes based electrodes", *Chemical Physics Letters*, v. 434, no. 1-3, 2007, pp. 73-77.

[9] Farahmandi et al., "Method of making a high performance ultracapacitor", *US Patent No. 6,059,847*, May 9, 2000.

[10] Gadkaree et al., "High energy density ultracapacitor", *US Patent Application No. 11/980,883,* April 30, 2009.

[11] Schindall et al., "Engineered structure for charge storage and method of making", *US Patent Application No. 11/429,565,* November 8, 2007.

[12] Ehrenberg et al., "Nanoparticle ultracapacitor", *US Patent Application No. 11/879,482*, December 25, 2008.

[13] Gadkaree et al., "Ultracapacitor with improved aging performance", *US Patent No. 8,564,934*, Oct. 22, 2013.

[14] Ezzat, G. Bakhoum, "New Mega-Farad Ultracapacitors", *IEEE Transactions on Ultrasonics, Ferroelectrics, and Frequency Control*, v. 56, no. 1, 2009, pp. 14-21.

[15] Ezzat G. Bakhoum and Marvin H. M. Cheng, "Electrophoretic Coating of Carbon Nanotubes for High Energy-Density Capacitor Applications", *J. of Applied Physics*, v. 105, no. 10, 2009, pp. 104314(1-6).

[16] F. L. Luo and H. Ye, "Investigation of Luo-Ye Tron", *Proceedings of the 5th Conference on Industrial Electronics and Applications, ICIEA*, 2010, pp. 473-479.

[17] S. Chen et al., "Graphene Oxide – MnO2 Composites for Supercapacitors", *ACS Nano*, v. 4, no. 5, 2010, pp. 2822-2830.

[18] K. Wang et al., "Conducting Polyaniline Nanowire Arrays for High Performance Supercapacitors", *J. Phys. Chem.*, v. 114, no. 17, 2010, pp. 8062-8067.

[19] Q. Cheng et al., "Graphene and nanostructured MnO_2 composite electrodes for supercapacitors", *Carbon*, v. 49, 2011, pp. 2917-2925.

[20] Y. He et al., "Freestanding Three-Dimensional Graphene/MnO_2 Composite Networks As Ultralight and Flexible Supercapacitor Electrodes", *ACS Nano*, v. 7, no. 1, 2013, pp. 174-182.

[21] W. H. Hayt and J. A. Buck, "*Engineering Electromagnetics*", (McGraw Hill, New York, NY, 2006).

[22] G. Arlt, D. Hennings and G. deWith, Dielectric Properties of Fine-Grained Barium Titanate Ceramics, *J. Applied Physics*, v. 58, 4, 1985, pp. 1619-1625.

[23] T. Tsurumi, T. Hoshina, H. Takeda, Y. Mizuno, and H. Chazono, "Size Effect of Barium Titanate and Computer-Aided Design of Multilayered Ceramic Capacitors", *IEEE Trans. on Ultrasonics, Ferroelectrics, and Frequency Control,* v. 56, no. 8, 2009, pp. 1513-1522.

[24] TPL, Inc. "Ceramics and Nano-Composites", www.tplinc.com.

[25] S. L. Swartz, T. R. Shrout, W. A. Schulze, L. E. Cross, "Dielectric Properties of Lead Magnesium Niobate Ceramics", *J. of the American Ceramic Society*, v. 67, no. 5, 1984, pp. 311-314.

[26] B. A. Tuttle, J. A. Voigt, D. L. Sipola, W. R. Olson, and D. M. Goy, "Chemically Prepared Lead Magnesium Niobate Dielectrics", *1997 MRS Fall Symposium,* v. 495, Article no. 236857.

[27] S. Haber and L. Gal-Or, "Deep Electrophoretic Penetration and Deposition of Ceramic Particles Inside Porous Substrates - Part I", *J. Electrochemical Society*, v. 139, no. 4, 1992, pp. 1071-1078.

[28] L. Gal-Or, S. Liubovich and S. Haber, "Deep Electrophoretic Penetration and Deposition of Ceramic Particles Inside Porous Substrates - Part II", *J. Electrochemical Society*, v. 139, no. 4, 1992, pp. 1078-1081.

[29] S. Haber, "Deep Electrophoretic Penctration and Deposition of Ceramic Particles Inside Impermeable Porous Substrates", *J. of Colloid and Interface Science,* v. 179, 1996, pp. 380-390.

[30] O. O. Van der Biest and L. J. Vandeperre, "Electrophoretic Deposition of Materials", *Annu. Rev. Mater. Sci.*, v. 29, 1999, pp. 327-352.

[31] E. Matijevic and M. Borkovec, "*Surface and Colloid Science*", (Wiley, New York, NY, 1969).

[32] *US Dept. of Energy Report No. DE-FG02-96ER82 138,* "Solid State Ultracapacitors for Electric Vehicles and Consumer Electronics", June 1999.

[33] M. D. Stoller, S. Park, Y. Zhu, J. An, and R. S. Ruoff, "Graphene-Based Ultracapacitors", *Nano Letters*, v. 8, no. 10, 2008, pp. 3498-3502.

[34] J. Huang, B. G. Sumpter and V. Meunier, "A Universal Model for Nanoporous Carbon Supercapacitors Applicable to Diverse Pore Regimes, Carbon Materials, and Electrolytes", *Chemistry, a European Journal*, v. 14, 2008, pp. 6614-6626.

[35] E. Lust, A. Janes, T. Parn, and Priit Nigu, "Influence of nanoporous carbon electrode thickness on the electrochemical characteristics of a nanoporous carbon/tetraethylammonium tetrafluoroborate in acetonitrile solution interface", *J. of Solid State Electrochemistry*, v. 8, 2004, pp. 224-237.

[36] C. M. Yang et al., "Nanowindow-Regulated Specific Capacitance of Supercapacitor Electrodes of Single-Wall Carbon Nanohorns", *J. of the American Chemical Society,* v. 129, 2007, pp. 20 21.

[37] J. Muzikar, T. van de Goor, B. Gas, and E. Kenndler, "Electrophoretic mobilities of large organic ions in nonaqueous solvents: Determination by capillary electrophoresis in propylene carbonate, N,N-dimethylformamide, N,N,-dimethylacetamide, acetonitrile and methanol", *Electrophoresis*, v. 23, 2002, pp. 375-382.

[38] Y. J. Kim et al., "High energy-density capacitor based on ammonium salt type ionic liquids and their mixing effect by propylene carbonate",

Report by Shinshu University, Wakasato, Nagano, Japan, 2004 (unpublished).

[39] Y. Xue and M. A. Ratner, "Theoretical Principles of Single-Molecule Electronics: A Chemical and Mesoscopic View", *Int. J. of Quantum Chemistry,* 2004, online e-print archive: http://arxiv.org/abs/cond-mat/0410539.

[40] H. Y. Sun, Y. Takeda, N. Imanishi, O. Yamamoto, and H. J. Sohn, "Ferroelectric Materials as a Ceramic Filler in Solid Composite Polyethylene Oxide-Based Electrolytes", *J. of The Electrochemical Society*, v. 147, no. 7, 2000, pp. 2462-2467.

[41] B. Kumar, S. J. Rodrigues and L. G. Scanlon, "Ionic Conductivity of Polymer-Ceramic Composites", *J. of the Electrochemical Society*, v. 148, no. 10, 2001, pp. A1191-A1195.

[42] C. Capiglia et al., "Composite Polymer Electrolyte: the Role of Filler Grain Size", *Solid State Ionics*, v. 154-155, 2002, pp. 7-14.

[43] P. K. Singh and A. Chandra, "Role of the dielectric constant of ferroelectric ceramic in enhancing the ionic conductivity of a polymer electrolyte composite", *J. of Applied Physics D*, v. 36 , 2003, pp. L93-L96.

[44] K. Bullis, "A Trick for Making Batteries Safer Could Also Make EVs Affordable", *MIT Technology Review*, v. 116, no. 6, Nov. 2013, pp. 67-68.

[45] Dennis M. Zogbi, "Supercapacitors: the Myth, the Potential and the Reality", *TTI Technology group report*, April 3, 2013, http://www.ttiinc.com.

[46] Market Research Media, *"Ultracapacitor Market Forecast 2015-2020"*, May 2013, http://www.marketresearchmedia.com/?p=912.

[47] Frost and Sullivan, *"World Ultracapacitor Markets"*, May 2013, http://www.frost.com.

[48] Innovative Research and Products Inc., *"Ultracapacitors for Stationary, Industrial, Consumer and Transport Energy Storage – An Industry, Technology and Market Analysis"*, Feb. 2010, http://www.innoresearch.net/report_summary.aspx?id=71andpg=171andrcd=ET-111andpd=2/1/2010.

[49] C. Ahern, "Supercapacitors", *Foresight Science and Technology report*, Dec. 2009, p. 2.

[50] Burke, "Ultracapacitor Technologies and Application in Hybrid and Electric Vehicles", *International Journal of Energy Research*, Jul. 2009, pp. 1-29.

[51] Norton, D. P., "Synthesis and Properties of Epitaxial Electronic Oxide Thin-Film Materials", *Materials Science and Engineering R-reports*, v. 43, 5-6, 2004, pp. 139-247.

[52] Z. J. Zhang, S. Fan, J. Huang, and C. M. Lieber, "Pulsed laser deposition and physical properties of carbon nitride thin films", *J. of Electronic Materials,* v. 25, no. 1, 1996, pp. 57-61.

[53] E. Broitman, N. Hellgren, J. Neidhardt, I. Brunell, and L. Hultman, "Electrical properties of Carbon Nitride thin films: role of morphology and Hydrogen content", *J. of Electronic Materials*, v. 31, no. 9, 2002, pp. 957-961.

[54] J. Gamby, P. L. Taberna, P. Simon, J. F. Fauvarque, and M. Chesneau, "Studies and characterisations of various activated carbons used for carbon/carbon supercapacitors", *J. of Power Sources*, v. 101, no. 1, 2001, pp. 109-116.

[55] Leu, L. C., Sadik, P., Norton, D. P., McElwee-White, L., Anderson, T. J., "Comparative Study of ZrN and Zr-Ge-N Thin Films as Diffusion Barriers for Cu Metallization on Si", *Journal of Vacuum Science and Technology B,* v. 26, 2008, pp. 1723-1727.

[56] Leu, L. C., Norton, D. P., McElwee-White, L., Anderson, T. J., "Properties of Reactively Sputtered W-B-N Thin Film as a Diffusion Barrier for Cu Metallization on Si", *Applied Physics A-Materials Science and Processing,* v. 94, 2009, pp. 691-695.

[57] Leu, L. C., Norton, D. P., McElwee-White, L., Anderson, T. J., "Ir/TaN as a Bilayer Diffusion Barrier for Advanced Cu Interconnects", *Applied Physics Letters,* v. 92, no. 11, 2008, Art. No. 111917.

[58] M. Markku and M. Ritala, "Atomic Layer Deposition Chemistry: Recent Developments and Future Challenges", *Angewandte Chemie*, v. 42, no. 45, 2003, pp. 5548-5554.

[59] M. Wegmann, L. Watson and A. Hendry, "XPS Analysis of Submicrometer Barium Titanate Powder", *J. Am. Ceramic Society*, v. 87, 3, 2004, pp. 371-377.

[60] B. L. Cheng, B. Su, J. E. Holmes, T. W. Button, M. Gabbay, and G. Fantozzi, "Dielectric and Mechanical Losses in (Ba,Sr)TiO_3 Systems", *J. of Electroceramics,* v. 19, no. 1, 2002, pp. 17-23.

[61] M. W. Cole, S. Hirsch, E. Ngo, and C. Hubbard, "Design, fabrication and material properties of temperature stable performance consistent tunable devices", *Integrated Ferroelectrics*, v. 101, no. 1, 2008, pp. 182-194.

[62] Electronic Industries Alliance, *"EIA Temperature Coefficients: Ceramic Capacitors"*, 2002, http://www.eia.org.

[63] L. D. Woolf et al., "Continuous fabrication of high-temperature superconductor coated metal fiber and multifilamentary wire", *Applied Physics Letters,* v. 58, no. 5, 1991, pp. 534-536.

[64] T. Ishihara, K. Sato and Y. Takita, "Electrophoretic Deposition of Y_2O_3-Stabilized ZrO_2 Electrolyte Films in Solid Oxide Fuel Cells", *J. of the American Ceramic Society,* v. 79, no. 4, 1996, pp. 913-919.

[65] A. R. Boccaccini, O. Van der Biest and J. B. Talbot, "*Electrophoretic Deposition – Fundamentals and Applications*", (Electrochemical Society, Pennington, NJ, 2002).

[66] H. W. Wang et al., "Formation of Barium Strontium Titanate Thin Films via Electrophoretic Deposition Process", *IEEE Trans. on Ultrasonics, Ferroelectrics, and Frequency Control,* v. 55, no. 12, 2008, pp. 2539-2543.

[67] S. Doungdaw et al., "Electrophoretic deposition of lead zirconate titanate (PZT) powder from ethanol suspension prepared with phosphate ester", *Science and Technology of Advanced Materials*, v. 6, 2005, pp. 927–932.

[68] N. Biskup, A. D. Andres, J. L. Martinez, and C. Perca, "Origin of the Colossal Dielectric Response of $PrCaMnO_3$", *Phys. Rev. B,* v. 72, 2, 2005, Art. No. 024115.

[69] S. Guillement-Fritsch et al., "Colossal Permittivity in Ultrafine Grain Size $BaTiO_3$ and $BaLaTiO_3$ Materials, *Adv. Materials*, v. 20, 2008, pp. 551-555.

[70] Z. Valdez-Nava et al., "Colossal Dielectric Permittivity of $BaTiO_3$ -based Nanocrystalline Ceramics Sintered by Spark Plasma Sintering", *J. Electroceramics,* v. 22, 2009, pp. 238-244.

[71] C. Romero and S. Baldelli, "Sum Frequency Generation Study of the Room-Temperature Ionic Liquids-Quartz Interface", *J. of Physical Chemistry B*, v. 2006, no. 110, 2006, pp. 6213-6223.

[72] C. Aliaga and S. Baldelli, "A Sum Frequency Generation Study of the Room-Temperature Ionic Liquid-Titanium Dioxide Interface", *J. of Physical Chemistry C,* v. 2008, no. 112, 2008, pp. 3064-3072.

[73] S. Baldelli, "Surface Structure at the Ionic Liquid-Electrified Metal Interface", *Accounts of Chemical Research*, v. 41, no. 3, 2008, pp. 421-431.

[74] T. Devarajan et al., "Novel ionic liquid electrolyte for electrochemical double layer capacitors", *Electrochemistry Communications*, v. 11, 2009, pp. 680-683.

[75] P. Hohenberg and W. Kohn, "Inhomogeneous Electron Gas", *Phys. Rev.*, 136, 1964, pp. B864-B871.

[76] W. Kohn and L. J. Sham, "Self-Consistent Equations Including Exchange and Correlation Effects", *Phys. Rev.*, 140, 4, 1965, pp. 1133-1138.

[77] R. E. Cohen and H. Krakauer, "Lattice Dynamics and Origin of Ferroelectricity in BaTiO3: Linearized-Augmented Plane Wave Total Energy Calculations", *Phs. Rev. B,* 42, 10, 1990, pp. 6416-6423.

[78] R. D. King-Smith and D. Vanderbilt, "Theory of Polarization of Crystalline Solids", *Phys. Rev. B*, 47, 3, 1993, pp. 1651-1654.

[79] D. Vanderbilt and R. D. King-Smith, "Electric Polarization as a Bulk Quantity and its Relation to Surface Charge", *Phys. Rev. B*, 48, 7, 1993, pp. 4442-4455.

[80] W. Zhong, D. Vanderbilt and K. M. Rabe, "Phase Transitions in BaTiO3 from First Principles", *Phys. Rev. Lett.*, 73, 13, 1994, pp. 1861-1865.

[81] R. D. King-Smith and D. Vanderbilt, "First-Principles Investigation of Ferroelectricity in Perovskite Compounds", *Phys. Rev. B,* 49, 9, 1994, pp. 5828-5844.

[82] R. W. Nunes and D. Vanderbilt, "Real-Space Approach to Calculation of Electric Polarization and Dielectric Constants", *Pys. Rev. Lett.*, 73, 5, 1994, pp. 712-715.

[83] N. Sai, K. M. Rabe and D. Vanderbilt, "Theory of Structural Response to Macroscopic Electric Fields in Ferroelectric Systems", *Phys. Rev. B*, 66, 2002, pp. 104108(1)-(17).

[84] O. Dieguez and D. Vanderbilt, "First-Principles Calculations for Insulators at Constant Polarization", *Phys. Rev. Lett.,* 96, 2006, pp. 056401(1)-(4).

[85] R. Resta and D. Vanderbilt, *"Theory of Polarization: A Modern Approach"* in Physics of Ferroelectrics: a Modern Perspective, ed. by K. M. Rabe, C. H. Ahn and J. M. Triscone, Springer-Verlag, Berlin, 2007, pp. 31-68.

[86] X. Gonze et al., *"ABINIT: First Principles Studies of Materials"*, www.abinit.org, 2008.

[87] W. J. Hehre, L. Radom and J. A. Pople, "*Ab Initio Molecular Orbital Theory*", (Wiley, New York, NY, 1986).

[88] D. Bahadur, S. Vitta and O. Prakash, "*Inorganic Materials – Recent Advances*", (Alpha Science International, Ltd., Oxford, UK, 2004).

[89] Y. Gohda, Y. Nakamura, K. Watanabe, and S. Watanabe, "Self-Consistent Density Functional Calculation of Field Emission Currents from Metals", *Phys. Rev. Lett.,* v. 85, no. 8, 2000, pp. 1750-1753.

[90] M. Tanaka, Y. Gohda, S. Furuya, and S. Watanabe, "Ab Initio Calculation of Capacitance of Semi-Infinite Jellium Electrodes with a Nanoscale Gap", *Japanese J. of Applied Physics*, v. 42, no. 7A, 2003, pp. L766-L768.

[91] E. Heifets et al., "Ab initio calculations of the SrTiO3 (110) polar surface", *Physical Review B*, v. 69, 2004, Article No. 035408.

In: Hybrid Vehicles
Editor: Hilda Bridges
ISBN: 978-1-63482-157-5

Chapter 2

ANALYSIS OF HYBRID VEHICLE CONFIGURATIONS BASED ON REAL-WORLD ON-ROAD MEASUREMENTS

Gonçalo Duarte[1,*] and Patrícia Baptista[1]
[1]LAETA, IDMEC, Instituto Superior Técnico,
Universidade de Lisboa, Lisboa, Portugal

ABSTRACT

Hybrid vehicles are becoming increasingly available in the market, emphasizing the importance of a better understanding of its benefits in different driving conditions. Consumers have a distinct variety of hybrid designs available and this work intends to explore the differences between the two hybrid vehicle configurations (parallel/series and parallel configurations), based on a total of over 13 hours of 1 Hz real-world monitoring data. Five vehicles were monitored on-road and under real-world driving conditions, in Lisbon (Portugal). The vehicles were monitored with a Portable Emission Measurement System to collect second-by-second information on engine parameters, tailpipe emissions and road topography. The data collected was analyzed using the Vehicle Specific Power (VSP) methodology to perform an energy and environmental characterization of the vehicles.

---

* Corresponding author. Tel.: +351 218 419 554; fax: +351 218 417 640. Email address: goncalo.duarte@tecnico.ulisboa.pt.

The parallel/series configurations present lower fuel consumption for lower VSP modes, while the parallel configurations are more efficient for higher VSP modes. While parallel/series configuration can only use the electric motor to move the vehicle under low power conditions (up to 11 to 12 W/kg depending on the vehicle) and turn the ICE off during a considerable amount of the time spent on braking, deceleration and idling, the parallel configuration only turns the ICE off at idling and only in a small part of the braking and deceleration time. However, the electric motors are used to assist the ICE under higher power conditions (such as accelerations and hard starts). Therefore, these hybrid configurations present a trade-off, where the parallel/series configuration aims at reducing liquid fuel use mostly at low power conditions, while parallel configuration aims to reducing the liquid fuel use under high power conditions. Consequently, the energy and environmental performance of these vehicles is very dependent on the driving context.

Parallel/series hybrids present the lowest fuel consumption for the urban cycle, presenting, on average -30% of fuel consumption compared with average energy use of parallel configurations. Regarding the extra-urban driving cycle, the results are vehicle dependent and there is not a clear trend concerning which hybrid design presents the best fuel economy. Under highway conditions, parallel configuration uses the electric motor to support the ICE under acceleration, presenting the lowest fuel consumption, circa 11% lower than the full hybrid configuration. These conclusions can be transposed for CO2 emission and were also quantified for HC and NOx.

Summarizing, this work emphasizes not only the real-world impacts of the different hybrid configurations available, but also how effective they perform under typical drive-cycles, with different characteristics.

Keywords: On-road measurement, vehicle specific power, hybrid vehicles, portable emission measurement system

INTRODUCTION

The transportation sector which includes road transportation, rail, air, water and pipeline that is used to the mobility of people and goods is responsible for almost 30% of the world's total energy consumption. This sector has considerable impacts in terms of greenhouse gases (GHG), such as CO_2 and N_2O, as well as pollutant emissions [1]. For instance, between 1997 and 2008, more than 20% of the population living in urban areas of European

Energy Agency member countries was exposed to concentration levels of PM_{10} and NO_2 higher than the local limit values [2].

As far as road transportation, European EURO and American Clean Air Act Standards have the purpose of regulating light duty vehicle pollutant emissions, namely carbon monoxide (CO), hydrocarbons (HC), nitrous oxides (NO_x) and particulate matter (PM). Vehicles are tested on chassis dynamometer, performing a predefined speed cycle (New European Driving Cycle – NEDC or Federal Test Procedure - FTP) and exhaust emissions are collected on bags during the cycle. Light-duty vehicles in Europe are homologated if the mass of each pollutant per kilometer collected during NEDC is below the EURO standards.

In order to fulfill the tighter pollutant emission standards, automotive industry has introduced several technologies to improve energy efficiency and reduce pollutant emissions. One of these examples has been the introduction of hybrid vehicles during the last decade, since they combine different technologies/powertrains and energy sources, reducing the vehicle environmental impacts in its utilization stage The most common hybrid technology available for consumers combines electric energy stored in batteries and chemical energy contained in fuel stored onboard. There are mainly three configurations of hybrid vehicles: parallel, series and parallel/series or full-hybrid:

- The parallel configuration uses a conventional spark-ignition internal combustion engine (ICE) as the main power supplier to the wheels. In commercially available parallel hybrids, the internal combustion engine is the main responsible for propulsion and the electric motor usually assists in standing starts and accelerations. The electric motor is usually located between the internal combustion engine and the transmission, providing a compact, lightweight and simple arrangement [3, 4]. This motor also acts as a generator when braking in order to charge the on-board batteries;
- In the serial configuration, energy is supplied by internal combustion engine which operates as a generator, supplying power to the electric motor and charging on-board batteries. Regenerative braking also allows charging the batteries. This configuration is more suitable for driving cycles involving many stop/starts in urban driving [5]. By not being directly connected to the wheels, the internal combustion engine can be downsized [6] and optimized to a narrow working range [7]; and

- The parallel/series configuration (full hybrid) allows for different propulsion strategies, according to the driving needs: electric assist, electric traction, or simply ICE power supply. Power control strategies have a tendency to provide at all times a maximum fuel economy as well as a minimum emission and also a good driving performance. The interaction between ICE and electric motors is made through a planetary gear (usually referred as power split device) connected to the electric motor and internal combustion engine [8]. The ICE works under defined conditions of high engine load and, consequently, low specific fuel consumption and high efficiency.

Moreover, under real-world on-road operation, the relations between internal combustion engine and electric motor are mostly dependent on the battery state-of-charge and driving power requirements [9]. The particularities of each propulsion system reinforce the importance of monitoring under real-world conditions to capture typical operation of these new vehicle technologies, using adequate tools to collect driving dynamics and energy management, as well as pollutant emission data. The portable emission monitoring system (PEMS) are used to perform real world on-road monitoring of vehicle technologies since they provide at second by second the characterization of engine parameters, tailpipe emissions and fuel consumption, which can be related with vehicle dynamics during on-road measurement. PEMS have been used by several authors [10, 11, 12] to characterize the energy and environmental impacts of light-duty, heavy-duty or even non-road vehicles.

Regarding data treatment from on-road monitoring, event-based analysis of second by second data is not feasible and data must be globally aggregated and analyzed. Thus, a road-load model based on the physics of the vehicle is used - the Vehicle Specific Power (VSP) methodology. VSP shows a good correlation between emissions and estimated power per mass unit [13] and evaluates the power per mass unit required for a given combination of speed, acceleration, road grade, aerodynamic and rolling resistance in a second-by-second basis and associates the correspondent mass of fuel used and pollutants emitted. According to the specific power, on-road data is grouped in bins, where each bin or mode has associated an average fuel consumption and correspondent mass of pollutants [14]. This allows relating vehicle dynamics and road grade with ICE variables and exhaust emissions to estimate fuel consumption, CO_2 and pollutant emissions as a function of VSP mode. As a consequence, it is possible to establish comparisons between different propulsion technologies independently of the driving cycle used.

Consequently, this work aims at analyzing a series of issues which are connected to the vehicle's behavior such as parallel and parallel/series hybrid systems under typical urban, urban/extra-urban and highway driving cycles in order to evaluate their impacts in terms of operation, energy and environment.

METHODOLOGY

This work was based on data from on-road measurements under urban and non-urban roads as well as the driving conditions at highways. Measurements were made in Lisbon, Portugal, comprehending diverse types of driving conditions, such as uphill/downhill as well as flat zones in urban and extra-urban areas The road test measurements were carried out with 5 hybrid vehicles provided by major automotive manufacturers, with a total monitoring period of over 13 hours and an average of 2.7 hours per vehicle.

PEMS DESCRIPTION

A PEMS was installed in the 5 monitored vehicles, the data collection was gathered on a second by second basis and the information on the vehicle dynamics, road topography, engine parameters and exhaust gases were recorded [9].

The collected data is provided by several equipment installed in the vehicle, which are connected to a laptop that runs the software developed in the LabView to integrate, synchronize and record all the information collected along the trip. The internal combustion engine parameters are obtained directly from a multi-protocol OBD port reader. Supported vehicle sensor data includes information on vehicle speed, engine speed and load, air flow mass, manifold absolute pressure, intake air temperature, throttle position and coolant temperature. A GPS receiver with barometric altimeter is used to collect latitude, longitude and altitude along the trip for posterior calculation of road grade. Tailpipe emissions were measured with a portable gas analyzer [9]. It provides simultaneous information about carbon dioxide (CO_2), carbon monoxide (CO), hydrocarbons (HC), nitrous oxides (NO_x) and oxygen (O_2). Non-dispersive infrared chambers are used to evaluate CO_2, CO and HC concentrations, while oxygen and NO_x are measured with electro-chemical sensors. The collected data allows estimating the ICE fuel consumption and the mass of pollutants present in the exhaust in a second by second basis.

Since liquid fuel is the only external source of supply, hybrid vehicles were considered as "black boxes" in powertrain management and under long-term operation, fuel consumption rates would tend to an average value, consequence of the vehicle electric/ICE powertrain management [9].

VEHICLE SPECIFIC POWER METHODOLOGY

An energy and environmental analysis was made through the use of the Vehicle Specific Power (VSP) methodology [10, 11, 13]. VSP allows estimating the power per mass unit necessary for a given driving conditions, based on a combination of vehicle dynamics (speed, acceleration, rolling and aerodynamic resistance) and road grade, as is presented in Eq. 1.

$$VSP = \frac{\frac{d}{dt}(E_{Kinetic} + E_{Potential}) + F_{Rolling} \cdot v + F_{Aerodynamic} \cdot v}{m} \quad (=)$$

$$(=) \frac{\frac{d}{dt}\left(\frac{1}{2}m \cdot (1+\varepsilon_i) \cdot v^2 + m \cdot g \cdot h\right) + C_r \cdot m \cdot g \cdot v + \frac{1}{2}\rho_a \cdot C_d \cdot A \cdot (v + v_w)^2 \cdot v}{m} \quad (=)$$

$$(=) v \cdot (a \cdot (1+\varepsilon_i) + g \cdot grade + g \cdot C_r) + \frac{1}{2}\rho_a \cdot \frac{C_d \cdot A}{m} \cdot (v + v_w)^2 \cdot v$$

Eq.1

where:

v: Vehicle speed (m/s);

a: Vehicle acceleration (m/s^2);

ε_i: Equivalent rotational mass of the rotating components of the powertrain, for manual transmission, depends on gear, hence the suffix *i*;

g: Gravitational acceleration;

$grade$: Road slope;

C_r: Rolling coefficient, which depends on road surface, tire type, vehicle speed, etc;

ρ_a: Ambient air density (kg/m^3);

C_d: Drag coeeficient, depends on vehicle configuration;

A: Vehicle frontal area (m^2);

v_w: Wind speed (m/s^2);
m: Vehicle mass (kg)

Simplifications were performed for the rolling resistance. The aerodynamic resistance terms and the effect of transmission gear used were based on data ranging from the compact vehicle to the luxury-sized one's [13]. Thus, at each point of the trip is given the correspondent VSP, according to Eq. 2.

$$VSP = v \cdot (1.1 \cdot a + 9.81 \cdot grade + 0.132) + 3.02 \cdot 10^{-4} \cdot v^3$$
Eq. 2

A modal analysis is performed where trip data points are grouped in bins or modes. By definition, for light-duty vehicles the VSP is usually divided in 14 modes and each one has statistically different fuel consumption values and none of them is dominant for the estimation of the trip total fuel consumption [11, 14].

Each VSP mode represents points of similar operation, where each mode has an energy footprint correspondent to the mass of fuel estimated based on engine and exhaust emissions data. The same approach is used for tailpipe emissions, where statistical information of pollutant emissions mass flow is provided for each VSP mode.

In order to obtain a higher level of definition in terms of the specific power demand, in this work the VSP was divided in modes of 1 W/kg, according to Table 1. This approach was first used in the MOVES model [14], before the establishment of the 14 modes.

Table 1. VSP binning and ranges of W/kg for each mode followed in this work

VSP Mode	W/kg	VSP Mode	W/kg	VSP Mode	W/kg	VSP Mode	W/kg	VSP Mode	W/kg	VSP Mode	W/kg
-21	VSP<-20	**-11**	-11≤VSP<-10	**-1**	-1≤VSP<0	**9**	8<VSP≤9	**19**	18<VSP≤19	**29**	28<VSP≤29
-20	-20≤VSP<-19	**-10**	-10≤VSP<-9	**0**	VSP=0	**10**	9<VSP≤10	**20**	19<VSP≤20	**30**	29<VSP≤30
-19	-19≤VSP<-18	**-9**	-9≤VSP<-8	**1**	0<VSP≤1	**11**	10<VSP≤11	**21**	20<VSP≤21	**31**	VSP>30
-18	-18≤VSP<-17	**-8**	-8≤VSP<-7	**2**	1<VSP≤2	**12**	11<VSP≤12	**22**	21<VSP≤22		
-17	-17≤VSP<-16	**-7**	-7≤VSP<-6	**3**	2<VSP≤3	**13**	12<VSP≤13	**23**	22<VSP≤23		
-16	-16≤VSP<-15	**-6**	-6≤VSP<-5	**4**	3<VSP≤4	**14**	13<VSP≤14	**24**	23<VSP≤24		
-15	-15≤VSP<-14	**-5**	-5≤VSP<-4	**5**	4<VSP≤5	**15**	14<VSP≤15	**25**	24<VSP≤25		
-14	-14≤VSP<-13	**-4**	-4≤VSP<-3	**6**	5<VSP≤6	**16**	15<VSP≤16	**26**	25<VSP≤26		
-13	-13≤VSP<-12	**-3**	-3≤VSP<-2	**7**	6<VSP≤7	**17**	16<VSP≤17	**27**	26<VSP≤27		
-12	-12≤VSP<-11	**-2**	-2≤VSP<-1	**8**	7<VSP≤8	**18**	17<VSP≤18	**28**	27<VSP≤28		

Test Drive Cycles

VSP modal analysis allows achieving an energy and environmental footprint, which is quantified by the mass flow of fuel, CO_2 and pollutant emission for a given VSP mode, independently of the driving cycle where the measurements were carried out. Also, each driving cycle has its own VSP time-based distribution, according to the speed, acceleration and road grade, as seen in Figure 1.

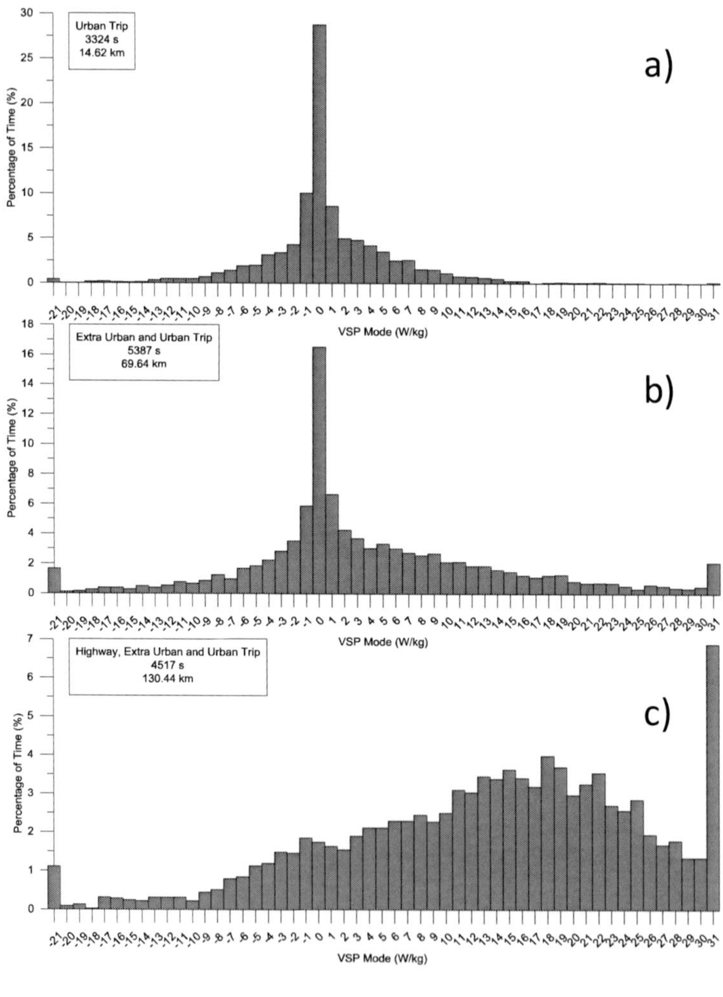

Figure 1. VSP time distributions for urban (a), extra-urban (b) and highway (c) driving conditions.

The trips considered for evaluating hybrid configuration performance on fuel use and emissions were collected under real-world operation to cover urban - Figure 1 a) -, extra-urban - Figure 1 b) - and highway - Figure 1 c) - driving conditions.

According with driving context, the VSP modal time distribution is different, with urban cycle dominated by low power demand and circa 30% of total time on idling; under extra-urban context (typically rural roads with little and moderate traffic) the VSP modal time distribution is more spread over the VSP modal range and circa 2% of time occurs in both extremes VSP modes (over 30 W/kg and below -20W/kg); for the monitored highway drive cycle, there is a shift towards the high power VSP modes, with the most extreme mode representing most of the time (circa 7% of total time).

In order to estimate the total fuel consumption for each trip, based on the vehicle characteristics and the time-distribution of the driving cycle [11], Equation 3 was used. A similar approach was used to estimate the pollutant emission. In that case, FC and FC_i were replaced by the correspondent pollutant.

$$FC = \sum_{i=1}^{14} FC_i \times t_i$$

Eq. 3

where:

FC: Total fuel consumption for a given driving cycle;

FC_i: Fuel consumption for the VSP mode *i* for a given vehicle;

t_i: time spent of VSP mode *I* for a given driving cycle.

Vehicle Characteristics

The tested vehicles include parallel and full hybrid configurations. The vehicles have been provided by the respective manufacturers. They were selected to cover most of the actual vehicle classes.

However, at the moment of the elaboration of this work, the vehicles with full hybrid configuration on the Supermini class and parallel configuration outside Supermini class were not available. The main characteristics of the vehicles are summarized in Table 2.

Table 2. Summary of the characteristics of the vehicles tested

Vehicle Class[1]	Supermini		Small family car		Large family car
Vehicle	A	B	C[2]	D[2]	E[2]
Technology	HEV-SI	HEV-SI	HEV-SI	HEV-SI	HEV-SI
Configuration	Parallel	Parallel	Full	Full	Full
ICE displacement (cc)	1496	1281	1798	1798	1798
ICE compression ratio	10.4:1	10.8:1	13.0:1	13.0:1	13.0:1
ICE Power (kW/RPM)	84/6000	64.9/5800	73/5200	73/5200	73/5200
ICE Torque (Nm/RPM)	145/4800	121/4500	142/4000	142/4000	142/4000
Vehicle mass (kg)	1130	1225	1495	1410	1725
Electric Power (kW/RPM)	10/1500	10.3/---	60/---	60/---	60/---
Electric Torque (Nm/RPM)	78/1000*	48.5/1000*	207/---	207/---	207/---
Combined Power (kW)	91	66	100	100	100
Total Power-to-Weight ratio (W/kg)	81	54	67	71	58
Measurements total time (s)	8069	10937	9463	4934	15051

[1] According to European New Car Assessment Programme (Euro NCAP).
[2] Vehicles R, S and Q share a similar powertrain.
* Constant up to 1000 rpm.

RESULTS

Characterization of on-Road Measurements

On-road vehicle monitoring data combined with the VSP methodology allows achieving an individual energy and environmental vehicle footprint. To assure that each VSP mode is covered by real-world measurements, it is necessary to guarantee that the routes monitored cover the widest possible range.

Figure 2 summarizes the time spent at each VSP mode during on-road measurements. During normal driving conditions it is difficult to achieve and maintain high power situations, consequently as VSP mode increases, the amount of time spent on those modes decreases. Thus, higher VSP modes are normally associated to accentuated road slope and high speed or high acceleration and speed or a combination of both.

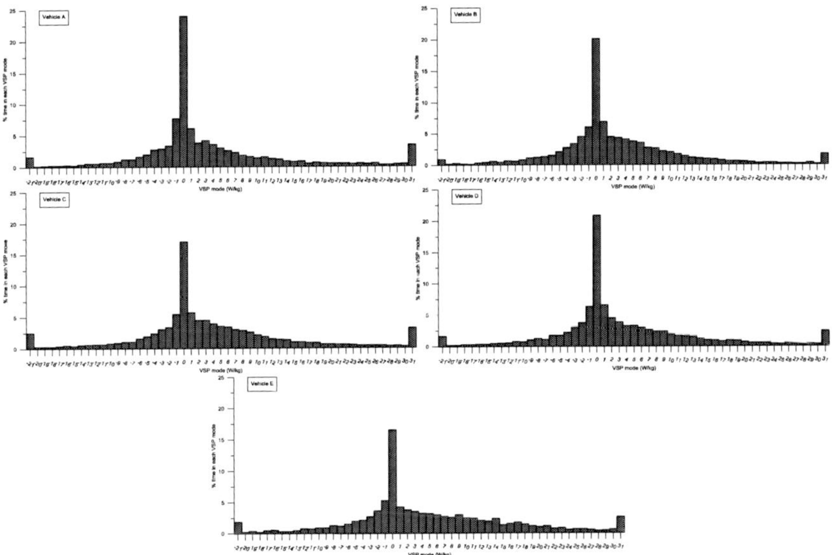

Figure 2. Time distribution for each VSP mode.

The routes where the measurements were performed have circa 80% of the total driving time in VSP mode lower or equal to 9 W/kg and ~90% of total time up to 16 W/kg. The last VSP modes congregate 10% of total time, where the highest VSP mode (31) comprehends all points over 30 W/kg, with an average time of 261 seconds in that mode (less than 3%), considering all vehicles.

Hybrid Strategies – ICE Management

Figure 3 presents the percentage of time where ICE was turned off for a given VSP mode. In Vehicles A and B (parallel configuration), ICE is turned off mainly at idling (VSP 0) and at light braking (lower negative VSP modes); however, these represent less than 15% of the total time on those modes. The differences found for ICE off in idling between Vehicle A and Vehicle B are possibly related with the fact that vehicle A is a compact sports car with manual gearbox and with an hybrid system designed for performance and with battery charge/discharge management that was not designed for fuel economy.

The parallel/series configuration (Vehicles C, D and E) presents considerable amounts of ICE off on negative VSP modes (circa 62%, on average).

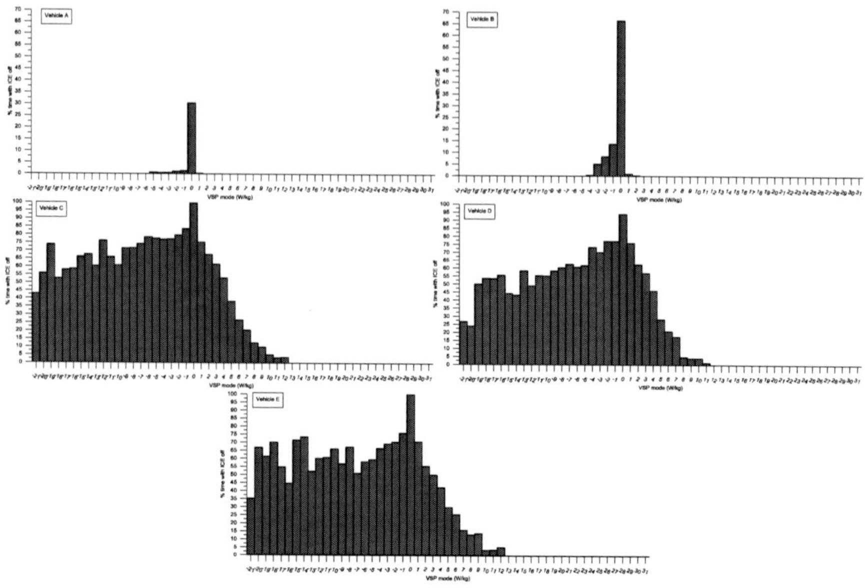

Figure 3. Percentage of time with ICE off, relative to the total time spent in a given VSP mode.

Since the full hybrid configurations allows the vehicle to be moved only by electric motors, using the energy stored in the battery, it is possible to observe ICE off up to VSP modes of 11-12 W/kg, according with the vehicle. After VSP mode 13, ICE is always turned on, either being assisted by the electric motor or providing all the necessary propulsion.

Figure 4 presents the average engine load for the hybrid configurations studied according with VSP mode. It is possible to conclude that full hybrid configuration uses the ICE at highest engine loads over the entire VSP range, in order to operate under conditions of high efficiency (low brake specific fuel consumption). It should be noted that, in this analysis, for the full hybrid configuration, only ICE on operation is included, in order to be comparable with parallel configuration, where ICE is always operating. For positive VSP modes, one can observe that the electric motor-ICE management is set in way that the ICE operates at high loads. As the lowest efficiency occurs at low loads, the electric motor is able to avoid these conditions. For parallel hybrids, the engine load increases monotonically with power which is quite similar to the way a conventional powertrain operates. However, and since the electric assistance occurs at standing starts and during acceleration, under these conditions, at higher VSP modes the slope of engine load increase is reduced, suggesting that part of the power is provided by the electric motor. For negative VSP modes, the full hybrid uses high engine loads in order to recharge the batteries when necessary.

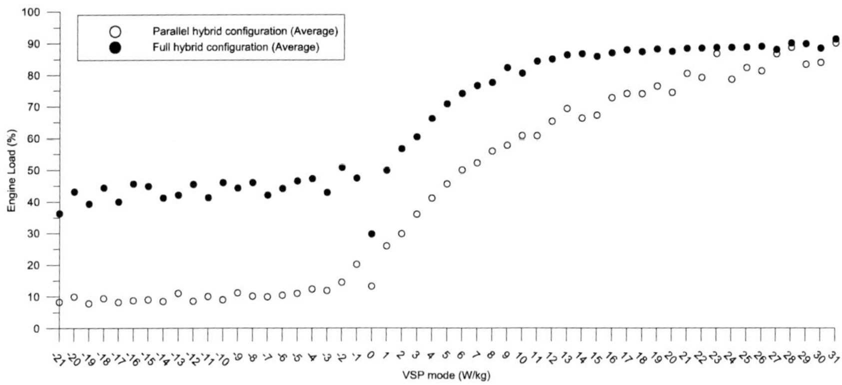

Figure 4. Internal combustion engine calculated load collected from vehicle OBD. Average values are presented for the parallel and full hybrid vehicles studied.

Energy and Environmental Footprint

In order to evaluate the energy consumption performance of each vehicle, the data collected from on-road monitoring with PEMS allows building fuel use and emission maps according to VSP. Figure 5 presents the fuel consumption rate for each VSP mode. As expected, fuel consumption rate increases with the power demand (higher VSP mode). On both parallel and full hybrid configurations it is possible to observe the impact of Stop/Start (at idling, VSP 0), presenting lower fuel consumption rates than the adjacent modes. On negative VSP modes, the fuel consumption rates are almost constant. Figure 6 presents the average fuel consumption rate per VSP mode, according with the hybrid configuration studied. The full hybrid configuration presents lower fuel consumption rates than the parallel configuration from modes 1 to 10 W/kg, where the sole electric motor propulsion occurs (confirmed also by the ICE off at these conditions). At higher VSP modes, the parallel hybrid configuration typically presents lower fuel consumption rates, due to the electric assist under acceleration. Moreover, the higher mass (circa more 200 kg) of the full hybrids studied also penalizes fuel consumption, particularly under high power (due to acceleration and speed, speed and road grade or a combination of all).

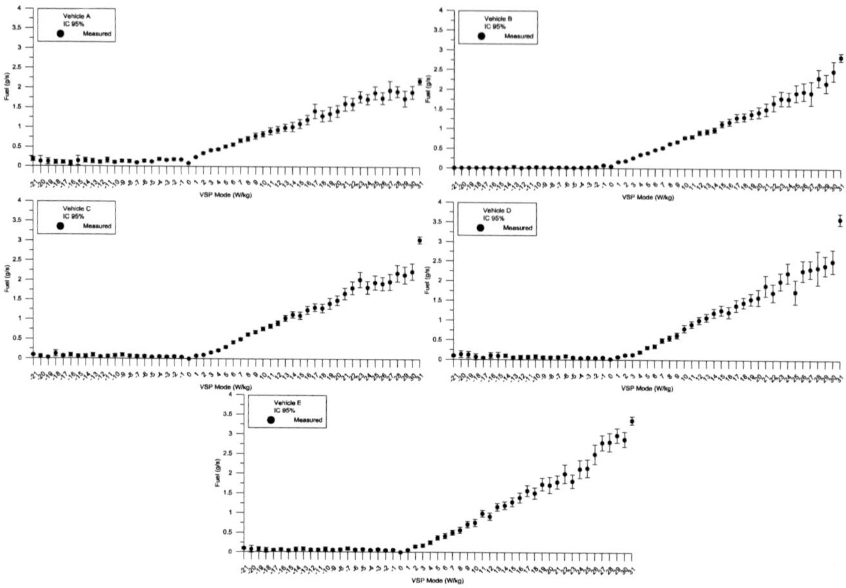

Figure 5. ICE fuel consumption (mass flow rate of gasoline) for each VSP mode.

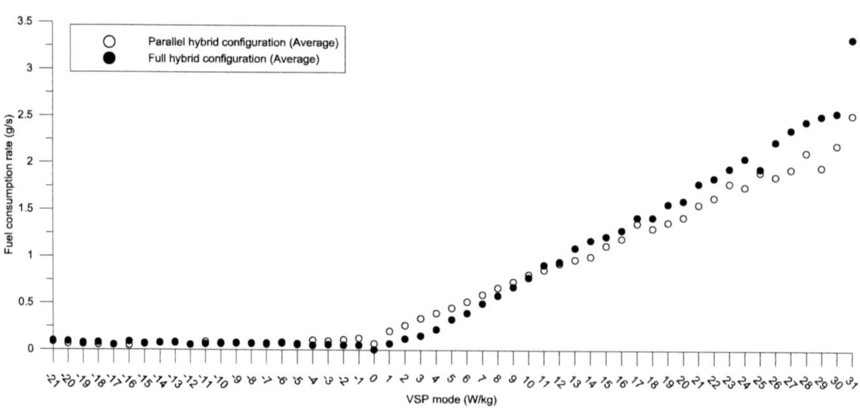

Figure 6. Average values of ICE fuel consumption according with hybrid configuration, per VSP modes.

Figure 7 presents the mass flow rate of CO_2 collected in the tailpipe during on-road measurements. Therefore, the CO_2 emission is directly related with the amount of fuel that is consumed, thus, CO_2 follows the same trend of fuel consumption. Figure 8 shows the mass flow of HC measured at the tailpipe for each VSP mode.

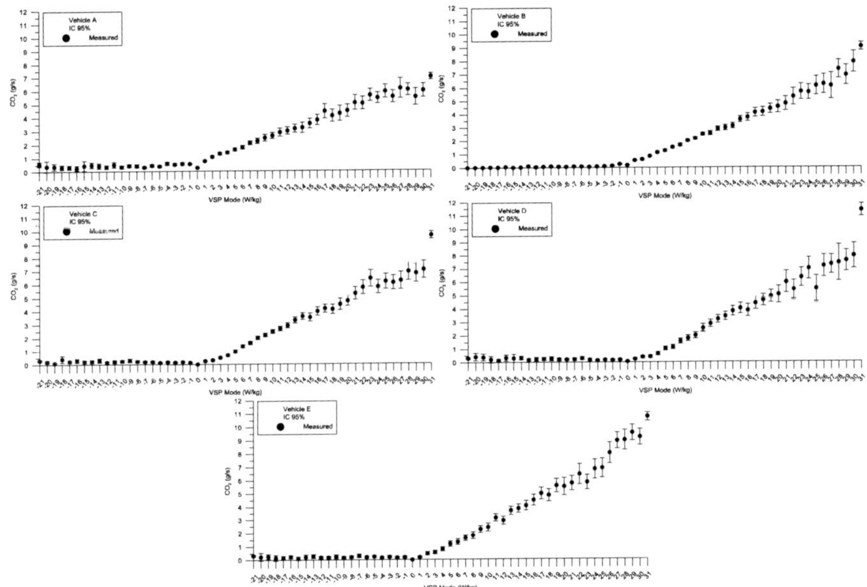

Figure 7. Exhaust mass emission rate of CO_2 for each VSP mode .

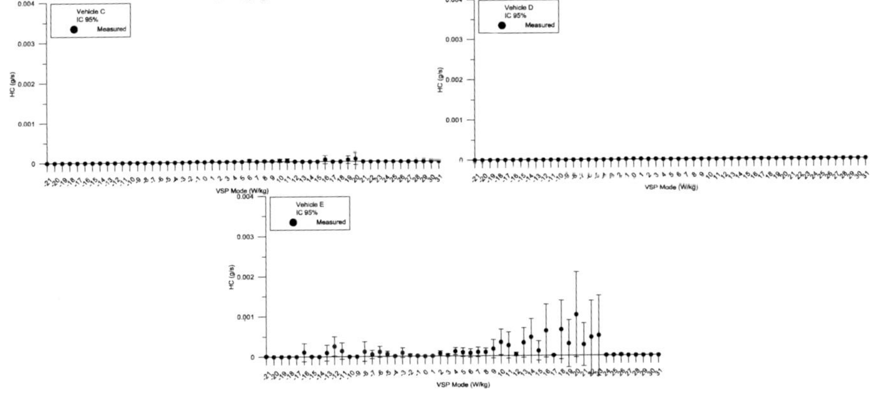

Figure 8. Exhaust mass emission rate of HC for each VSP mode .

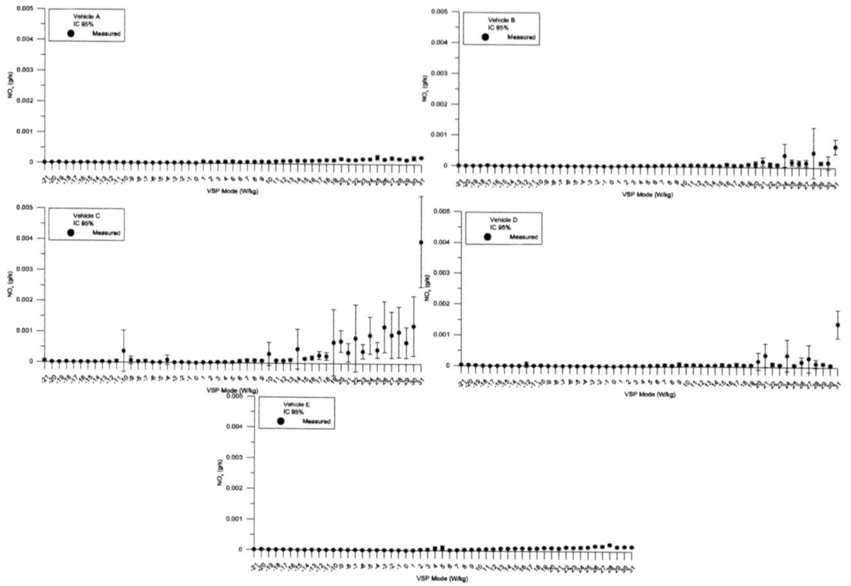

Figure 9. Exhaust mass emission rate of NO_x for each VSP mode.

The results achieved are very low, in the order of $10^{-3} - 10^{-4}$ g/s, which are near the detection limits of the analyzer used, but yet consistent with the high efficiency of the combustion of spark-ignition engines.

Figure 9 presents the mass emission rate of NO_X. The values are very low, ranging from 10^{-3} to 10^{-4} g/s. For spark ignition engines, NO_X emission is typically very low and depends not only on the burned gas temperature distribution but also in-cylinder pressure [15]. High VSP modes are normally associated to high loads as well as high RPM conditions, thus the tendency for NO_x to increase with VSP was verified in both hybrid configurations.

Energy and Environmental Performance on Urban and Urban/Extra-Urban Driving Cycles

Using either the formulation and the driving cycles presented in section 2.3 and, when combined with the VSP modal energy consumption and emission rates from section 3.3, it is possible to obtain not only the global fuel consumption but also the emission for the vehicles that have undergone the study of the specific driving cycles, with the final goal of evaluating the vehicle's performance in urban and extra-urban contexts.

Table 3. Estimated fuel consumption in tested driving cycles

	Vehicle A	Vehicle B	Vehicle C	Vehicle D	Vehicle E
	Urban				
Fuel consumption (l/100km)	8.9	6.3	5.1	5.3	5.5
IC 95% (l/100km)	0.2	0.1	0.1	0.2	0.2
	Extra-Urban				
Fuel consumption (l/100km)	5.6	5.0	4.9	5.2	5.5
IC 95% (l/100km)	0.1	0.1	0.1	0.1	0.1
	Highway				
Fuel consumption (l/100km)	5.2	5.3	5.4	5.9	6.2
IC 95% (l/100km)	0.1	0.1	0.1	0.1	0.1

Table 3 presents the fuel consumption data. The data shows that the parallel/series hybrid (vehicles C, D and E) have, on average, -30% of fuel consumption when compared to the average energy use of parallel configurations for the urban cycle.

The difference is circa -16% for the vehicle B and -40% for the vehicle A. Vehicle A, presents the highest fuel consumption on urban cycle since it is oriented for performance and not for fuel economy. Under urban context, the full hybrid propulsion system is very efficient for VSP modes 11 to 12, since the electric motor is used to move the vehicle, avoiding the use of the ICE under low load (and low efficiency) areas.

As far as this particular extra-urban driving cycle is concerned, it can be observed that the fuel consumption estimates are dependent from the vehicle which suggests that the vehicle characteristics are as important as the powertrain configuration. On the highway cycle, the full hybrid configuration is penalized at higher power conditions. These conditions can occur on the highway either by high speed, acceleration, road grade or by a combination of all this factors and, once the full hybrid configuration cannot use the electric motor by itself, the added weight can penalize them. Moreover, the parallel configuration uses the electric motor to support the ICE under acceleration, presenting the lowest fuel consumption, circa 11% lower than the full hybrid configuration.

The CO_2 data presented in Table 4 follow the same trends of fuel consumption, as it is expected, presenting a general decrease on CO_2 from urban cycle to extra-urban driving and highway.

Table 4. Estimated CO_2 emissions in tested driving cycles

	Vehicle A	Vehicle B	Vehicle C	Vehicle D	Vehicle E
	Urban				
CO_2 Emission (g/km)	204	146	118	124	128
IC 95% (g/km)	4	2	3	4	4
	Extra-Urban				
CO_2 Emission (g/km)	129	116	112	120	126
IC 95% (g/km)	2	2	2	2	2
	Highway				
CO_2 Emission (g/km)	120	123	126	137	144
IC 95% (g/km)	2	2	2	3	3

Table 5. Estimated HC emissions in tested driving cycles

	Vehicle A	Vehicle B	Vehicle C	Vehicle D	Vehicle E
	Urban				
HC Emission (g/km)	0.004	0.002	0.002	0.002	0.012
IC 95% (g/km)	0.001	0.000	0.000	0.000	0.003
	Extra-Urban				
HC Emission (g/km)	0.001	0.001	0.001	0.000	0.008
IC 95% (g/km)	0.000	0.000	0.000	0.000	0.002
	Highway				
HC Emission (g/km)	0.001	0.001	0.000	0.000	0.007
IC 95% (g/km)	0.000	0.000	0.000	0.000	0.003

Table 5 presents the HC emission for the vehicles and for the driving cycles studied. The results, as it was expected, were very low for spark-ignition engines, although the urban cycle presented the highest HC emission. As HC is typically higher for higher VSP modes (as it is shown in Figure 8), the low average speed can possible generate this effect. The results indicate that full hybrid configuration emits, on average, 78% more HC on the urban cycle, 200% on the extra-urban cycle and 133% on the highway cycle than the parallel configuration. In this sense, the ICE on/off operation of the full hybrid configuration can lead to higher emission outcomes [9].

Table 6. Estimated NO_x emissions in tested driving cycles

	Vehicle A	Vehicle B	Vehicle C	Vehicle D	Vehicle E
	Urban				
NO_x Emission (g/km)	0.009	0.004	0.010	0.004	0.005
IC 95% (g/km)	0.001	0.000	0.002	0.001	0.001
	Extra-Urban				
NO_x Emission (g/km)	0.005	0.004	0.015	0.005	0.003
IC 95% (g/km)	0.000	0.001	0.003	0.001	0.000
	Highway				
NO_x Emission (g/km)	0.004	0.004	0.020	0.006	0.003
IC 95% (g/km)	0.000	0.001	0.004	0.001	0.000

The data concerning NO_x are presented in Table 6, indicating that the NO_x emissions do not follow a trend with propulsion technology, but are closely related with the vehicle studied.

The higher NO_x emission of vehicle C in all the considered driving cycles is related with the high modal NO_x emission rates at the higher VSP modes. This is particularly clear in extra-urban and highway driving cycles. Regarding the urban driving cycle, the low average speed is responsible for the highest value presented (considering all vehicles). The results indicate that full hybrid configuration emits, on average, -7% less NO_x on the urban cycle than the parallel configuration. However, for the extra-urban cycle and the highway cycle, full hybrids emit more 70% and 142%, respectively, than the parallel hybrids.

CONCLUSION

This research work analyzes the commercially available hybrid vehicles in their different configurations (parallel and parallel/series hybrid vehicles) at typical urban, urban/extra-urban driving cycles, concerning their operational, energy and environmental impacts. A portable laboratory was used to perform on-road vehicle monitoring, collecting data at 1 Hz regarding engine operation, tailpipe gas composition, vehicle dynamics and road grade. Real-world operation data was analyzed using the VSP methodology to achieve an individual energy and environmental vehicle footprint.

The tested vehicles included parallel and full hybrid configurations, which were selected to cover most of the typical vehicle classes and have been provided by the respective manufacturers. In vehicles A and B (parallel configuration) the ICE is turned off mainly at idling (VSP 0) and at light braking (lower negative VSP modes), while the parallel/series configuration (vehicles C, D and E) presents considerable amounts of ICE off on negative VSP modes (over 60% of time, on average), as well as under low power demand. Since the full hybrid configurations allows the vehicle to be moved only by electric motors using the energy stored in the battery, it is possible to observe ICE off up to VSP modes of 11 to 12 W/kg. Moreover, when the ICE is operating, the full hybrid configuration uses the ICE at highest engine loads over the entire VSP range, in order to operate under conditions of high efficiency and low brake specific fuel consumption.

The modal fuel consumption collected under on-road conditions allows concluding that the full hybrid configuration presents lower fuel consumption rates than the parallel configuration on modes 1 to 10 W/kg, where the sole electric motor propulsion occurs, while after that power demand interval, the parallel configuration presents lower fuel consumption rates. The lowest vehicle weight (circa less 200 kg than full hybrids) and electric assistance under acceleration are likely to contribute to these results.

In order to evaluate the performance of these hybrid vehicles under different driving contexts, three generic urban, extra-urban and highway driving conditions collected under real-world conditions were selected to provide estimates of fuel use and emissions or the vehicles studied in these specific routes.

The results indicate that the parallel/series hybrids (vehicles C, D and E) present the lowest fuel consumption for the urban cycle, presenting, on average -30% of fuel consumption compared with average energy use of parallel configurations.

However, the difference is only circa -16% for vehicle B and -40% for vehicle A, since the hybrid management of this vehicle is adapted for performance, instead of fuel economy. Regarding the extra-urban driving cycle, the results are vehicle dependent and there is not a clear trend concerning which hybrid design presents the best fuel economy. Under highway conditions, parallel configuration uses the electric motor to support the ICE under acceleration, presenting the lowest fuel consumption, circa 11% lower than the full hybrid configuration. These conclusions can be transposed for CO_2 emission. Regarding pollutant emission, the results indicate that full hybrid configuration emits, on average, 78% more HC on the urban cycle, 200% on the extra-urban cycle and 133% on the highway cycle than the parallel configuration, although the HC emission is globally very low, as expected, for spark-ignition engines. For the NO_x the estimates obtained indicate that the full hybrid configuration emits, on average, -7% less NO_x on the urban cycle than the parallel configuration. However, for the extra-urban cycle and the highway cycle, full hybrids emit more 70% and 142%, respectively, than the parallel hybrids. The NO_x emission is also globally very low, accordingly with what was expected for vehicles with spark-ignition engines.

Summarizing, a detailed analysis has been performed on the energy and environmental performance of parallel and full hybrid vehicles based on real world on-road measurement, providing crucial information for the adequacy of alternative vehicle technologies to different driving contexts.

Submitted to:
Book entitled "Hybrid Vehicles: New Developments, Energy Management and Emerging Technologies"

October 2014

ACKNOWLEDGMENTS

Thanks are due to Fundação para a Ciência e Tecnologia for the Post-Doctoral financial support (SFRH/BPD/79684/2011). The authors would also like to acknowledge to vehicle manufacturers for providing the monitored vehicles.

REFERENCES

[1] U.S. Energy Information Administration. International Energy Outlook 2010, 2010. U.S. Department of Energy, Report nº DOE/EIA-0484 (2010).

[2] European Environment Agency. The European environment - state and outlook 2010: synthesis. 2010. European Environment Agency, DOI:10. 2800/45773.

[3] Honda. Technical Specifications. [Online] 2012. [Cited: 08 20, 2012.] http://world.honda.com/CR-Z/spec/index.html.

[4] Honda. The Honda Hybrid System. [Online] 2012. [Cited: 08 20, 2012.] http://world.honda.com/automobile-technology/IMA/ima03/.

[5] Momoh, O.; Omoigui, M., 2009. An Overview of Hybrid Electric Vehicle Technology. IEEE vehicle power and and propulsion conference.

[6] HILTech Developments Limited, 2006. HILTech Developments Limited vehicle hybrid powertrain architectures.

[7] Lo, E., 2009. Review on the Configurations of Hybrid Electric Vehicles. International conference power electronics systems and applications, ISBN: 978-1-4244-3845-7.

[8] Bayindir, K.; Gozukuçuk, M.; Teke, A., 2011. A comprehensive overview of hybrid electric vehicle: Powertrain configurations, powertrain control techniques and electronic control units. Energy Conversion and Management, 52, DOI: 10.1016/j.enconman.2010.09. 028.

[9] Duarte, G.; Varella, R.; Gonçalves, G.; Farias, T., 2014. Effect of battery state of charge on fuel use and pollutant emissions of a full hybrid electric light duty vehicle. *Journal of Power Sources,* 246, pp. 377-386, DOI: 10.1016/j.jpowsour.2013.07.103.

[10] Zhai, H.; Frey, H.; Rouphail, N., 2008. A Vehicle-Specific Power Approach to Speed- and Facility-Specific Emissions Estimates for Diesel Transit Buses. *Environmental Scientific Technology,* 42, DOI: 10. 1021/es800208d.

[11] Frey, H.; Rouphail, N.; Zhai, H.; Farias, T.; Gonçalves, G., 2007. Comparing real-world fuel consumption for diesel- and hydrogen-fueled transit buses and implication for emissions. *Transportation Research Part D,* 12, DOI: 10.1016/j.trd.2007.03.003.

[12] de Vlieger, I., 1997. On-Board Emission and Fuel Consumption Measurement Campaign on Petrol-Driven Passenger Cars. Atmospheric Environment, 31, pp. 3753-3761, DOI: 10.1016/S1352-2310(97)00212-4.

[13] Jiménez-Palacios, J., 1999. Understanding and Quantifying Motor Vehicle Emissions with Vehicle Specific Power and TILDAS Remote Sensing. PhD Thesis, Massachusets Institute of Technology.

[14] EPA - US Environmental Protection Agency, 2002. Methodology for Developing Modal Emission Rates for EPA's Multi-Scale Motor Vehicle and Equipment Emission System. Office of Transportation and Air Quality, U.S. Environmental Protection Agency, Report EPA420-R-02-027.

[15] Heywood, J., 1988. Internal Combustion Engine Fundamentals. McGraw-Hill, ISBN: 0-07-028637-X.

In: Hybrid Vehicles
Editor: Hilda Bridges

ISBN: 978-1-63482-157-5

Chapter 3

EMERGING ADVANCED PERMANENT-MAGNET BRUSHLESS MACHINES FOR HYBRID VEHICLES

Chunhua Liu and Wenlong Li *
Department of Electrical and Electronic Engineering,
The University of Hong Kong, Hong Kong, China

ABSTRACT

In this chapter, three emerging advanced permanent-magnet brushless machines are presented for hybrid vehicles. By introducing different types of hybrid vehicles, the power management for these vehicles is briefly introduced. Then, based on the aforementioned hybrid vehicle types, three emerging machines are presented for application in these vehicles, namely an outer-rotor permanent-magnet vernier motor for electric vehicle in-wheel motor drive, a dual-rotor dual-stator magnetic-geared PM machine for power splitting in series-parallel-type vehicles, and a hybrid stator-PM machine using as an integrated-starter-generator for complex-hybrid-type vehicles. The design equations, operating principle and analytical results of these machines are quantitatively elaborated and discussed.

* E-mail address: wlli@eee.hku.hk

INTRODUCTION

Due to ever increasing concerns on energy conservation and environmental protection, the hybrid vehicle (HV) is a widely accepted interim solution for evolving from the conventional internal combustion engine (ICE) vehicle to the clean electrified vehicle. Based on the drive train power management, HVs can be classified into 4 kinds, namely series hybrid, parallel hybrid, series-parallel hybrid and complex hybrid [1]. This chapter will emphasis on the newly developed permanent-magnet brushless (PMBL) machines utilizing as propulsion motors and generators for HVs.

This chapter consists of 5 sections and introduces 3 types of newly developed advanced PM motor for HVs. The origination of this chapter is as following:

Firstly, power management strategies of HVs are overviewed and compared. Power flow of series hybrid and parallel hybrid is relatively simpler than the other two kinds. Therefore, the series-parallel hybrid and complex hybrid are mainly emphasized in this chapter.

Secondly, a new type of outer-rotor in-wheel PM vernier motor is presented and analyzed for direct-drive propulsion which can be applied in the aforementioned 4 kind HVs. This kind of motor performs a low-speed operation using the magnetic gear effect. Therefore, the torque density and efficiency are improved. Both steady and dynamic performances will be presented.

Thirdly, another type of dual-rotor dual-stator magnetic-geared PM machine is proposed for series-parallel-type HVs. By connecting one rotor to the engine and coupling the other rotor to the transmission, this kind of motor can perform power splitting of the ICE output via two power converters as the planetary-gear electric continuous variable transmission in the Toyota Pruis. This machine utilizes the magnetic gearing effect and can realize the non-contact torque transmission and speed variation. Therefore, high transmission efficiency, silent operation and maintenance free can be achieved.

Fourthly, a type of hybrid stator-PM machine is designed and analyzed for motor/generator dual-mode operation. It is preferred for the complex-hybrid-type HVs. With flux strengthening, the high torque can be developed for electric launch or cold cranking. With flux weakening, the constant power range of this kind of motor can be significantly extended. With the aid of flexible flux control, the output voltage is kept constant with the variable engine speed when operating as a generator.

Finally, summary and conclusion of the 3 presented PM machines are presented.

POWER MANAGEMENT FOR HYBRID VEHICLES

Except for internal combustion engine (ICE), there are a lots of energy sources or stores, namely battery, fuel cell, ultra-capacitor, and flywheel can be adopted for the hybrid vehicles (HVs) [2]. Due to two or more energy sources involved in HVs, the power management system which coordinates and control power flow between sources becomes essential.

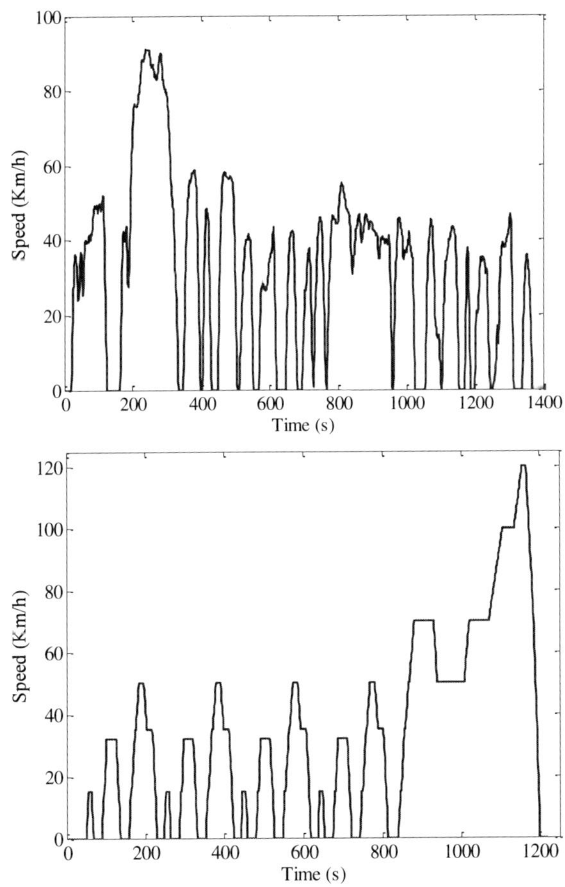

Figure 1. Driving cycles. (a) FTP-75. (b) NEDC.

Figure 1 shows the typical driving cycles: the EPA Federal Test Procedure (FTP-75) and the New European Driving Cycle (NEDC) [3]. FTP-75 illustrates the vehicle speed versus time during the urban driving while NEDC depicts the vehicle speed versus time during the urban driving and the highway driving. For the urban driving, the vehicle usually experiences a slow acceleration, constant speed cruising, slow deceleration and till a full stop. For the high way driving, except the above scenario, the high speed cruising time is usually longer than during the urban driving. Therefore, due to the diversity of vehicle states, a power management strategy is necessary to control and coordinate the power flow between multiple sources and the load. Based on different forms of coordination between the multiple energy sources in HVs, usually an ICE and one or several electric motors, HVs are classified into four groups, namely series hybrid, parallel hybrid, series-parallel hybrid and complex hybrid [1]. Power flows for the four types of HVs are presented and elaborated in the following sections.

SERIES HYBRID

The series hybrid vehicle is also termed as range-extended electric vehicles (REEV). A typical example is Chevrolet Volt made by General Motors [4]. In a series hybrid vehicle, the electric motor other than ICE is directly coupled to the mechanical transmission and the ICE is used to drive an electric generator to power the electric motor which is the only propulsion source for this vehicle.

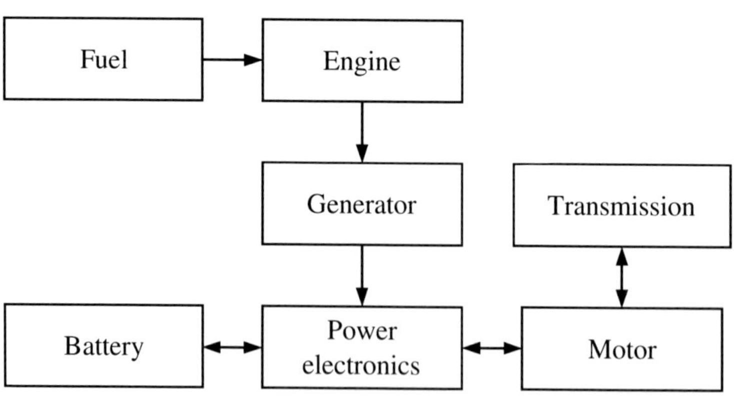

Figure 2. Possible power flows in a series hybrid system.

Figure 2 shows the possible power flows in a series hybrid system. During acceleration and constant speed cruising, both the ICE and the battery provide electrical energy to the electric motor (via the power electronics unit) which is directly coupled to the transmission. During deceleration, the vehicle kinetic energy can be retrieved by the electric motor which operates in its generator mode, and then charges the battery. For REEV, the electric energy from the battery is its main source. When the battery is depleted below the setting point, the ICE is started for charging the battery.

Parallel Hybrid

The parallel hybrid vehicle is a kind of HVs in which an ICE and an electric motor are coupled to the same axle to drive the vehicle. The two driving sources are independent to each other. A typical example is Honda Insight [5]. Figure 3 illustrates the possible power flows in a parallel hybrid system. For acceleration and constant speed cruising, both the ICE and electric motor operate together to drive the vehicle. For light load driving, one of the energy sources such as the ICE or the electric motor is turned off, and the rest source solely propel the vehicle. The selection of driving sources depends on how to share the motive power portion. For deceleration, the electric motor operate in its regenerative braking mode to retrieve the electric energy for charging the battery. Since the ICE and electric motor share the same shaft, the battery can be also charged by the engine via the electric motor.

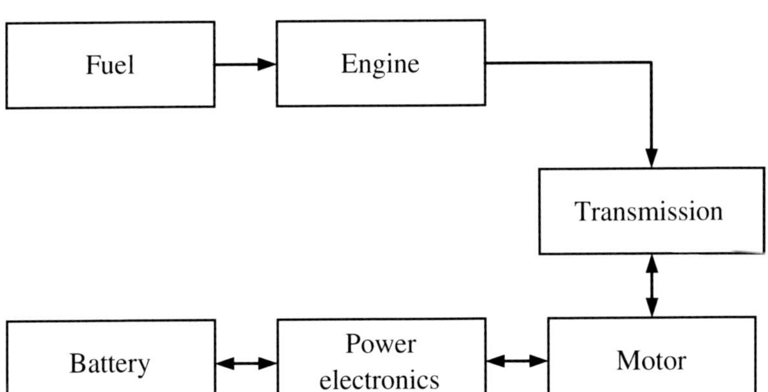

Figure 3. Possible power flows in a parallel hybrid system.

Series-Parallel Hybrid

The series-parallel hybrid vehicle possesses features of both series hybrid and parallel hybrid. Therefore, there are many possible power flow patterns in this type of HVs. A typical example is Toyota Prius [6]. Figure 4 illustrates the possible power flows in a series-parallel hybrid system. Since both the ICE and the electric motor are connected to the transmission directly, based on the power portion provided by the two sources, this type of HVs is further classified into the engine-heavy type and the electric-heavy type [2]. During the start-up, only electric drive is involved which is produced by the electric motor powered the battery. During the acceleration, the ICE and electric motor operate together to drive the vehicle for both type: for the electric-heavy type, the electric energy fed into the electric motor is provided by the battery and the ICE while for the engine-heavy type, the electric energy is provided by the battery only. During the light load driving, for the engine-heavy type, the ICE operates solely to propel the vehicle and the electric motor is turned off; whereas for the electric-heavy type, the ICE and the electric motor operate together to drive the vehicle and the electric energy fed into the electric motor is provided by the ICE. For deceleration, the electric motor operate in its regenerative braking mode to retrieve the electric energy for charging the battery. For battery charging mode, since the ICE and electric motor share the same shaft, the battery can be also charged by the engine via the generator.

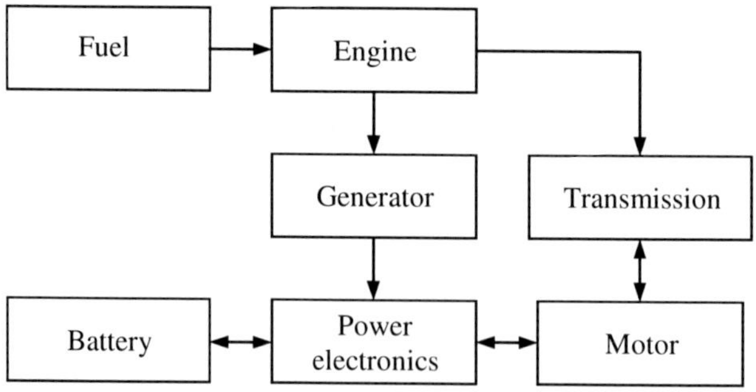

Figure 4. Possible power flows in a series-parallel hybrid system.

Complex Hybrid

There is a type of HVs which realize a four-wheel drive. In this system, the front wheel axle and the rear wheel axle have no direct connection and can be independently propelled. The two wheel axles can be driven by hybrid drive train and electric drive train separately. A typical example is General Motors Precept [7]. Figure 5 presents the possible power flows in this complex hybrid system. During the vehicle startup, for the front-hybrid and rear-electric type, the battery provides energy to both the front and rear electric motors to drive the vehicle and the front electric motor is served as a starter motor to starting the ICE and the rear electric motor propel the vehicle directly, while for the front-electric and rear hybrid type, only the front electric motor operates to propel the vehicle and the ICE is in off mode. During the acceleration, both of two types work in the same way: the ICE and one of electric motor operate together to drive one axle, and meanwhile the other electric motor also propels the other axle. During the constant speed cruising, for both types, the ICE works solely to propel the vehicle. For deceleration, both the front and rear electric motors operate in generator mode to charge to battery. For battery charging mode, the ICE drives the starter motor which works as a generator to charge the battery.

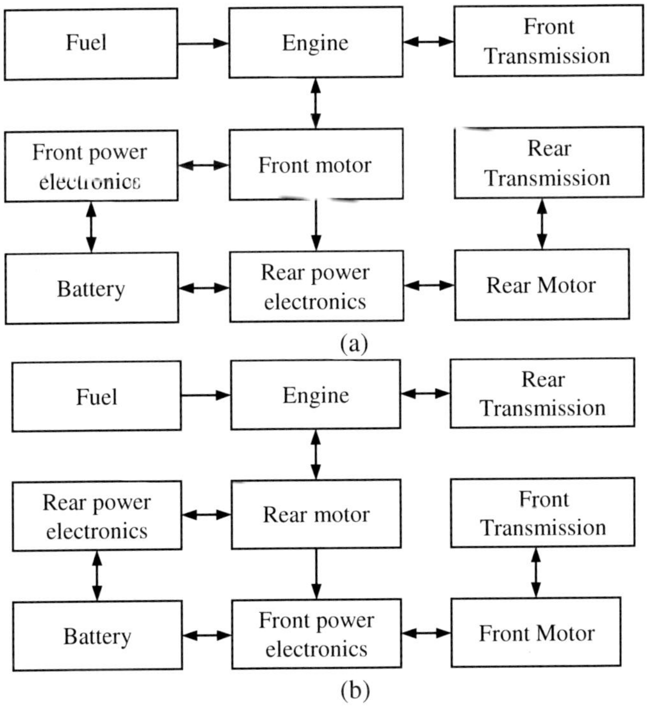

Figure 5. Possible power flows in a complex hybrid system. (a) Front-hybrid and rear-electric type. (b) Front-electric and rear-hybrid type.

PERMANENT-MAGNET VERNIER MOTOR FOR EV IN-WHEEL DRIVE

The permanent-magnet vernier (PMV) machine is a member of the variable reluctance permanent-magnet machine family [8]. Its magnetic circuit is featured as the slotted structure and multi-pole PMs [9]-[14]. As shown in Figure 6, it can be designed as a toothed-pole stator with PMs mounted on its rotor, and a stator with PM mounted on its tooth surface and a slotted rotor. The first one operates due to the two rotating magnetic field, and the second one works as a flux-reversal machine. A small movement of the rotor can cause a large flux-linkage variation in the armature winding which further results in a high torque. This is also known as the magnetic gearing effect which results from the interaction between the PMs and toothed-pole structure. Due to features of the high torque density and the compact structure, it is very suitable for the EV in-wheel drive.

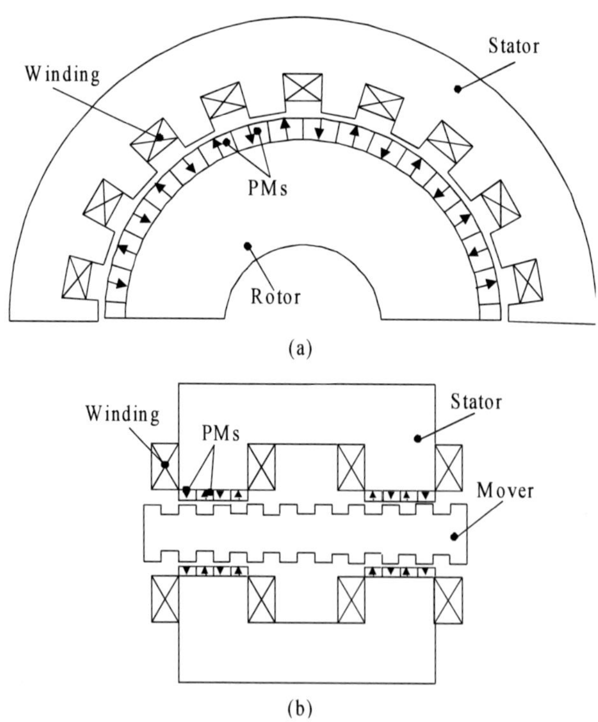

Figure 6. PMV machine. (a) Rotational morphology. (b) Linear morphology.

In this section, design procedure, analysis, optimization of the PMV machines studied. By studying its operation principle, its design procedure is introduced. Then, by analytically expressing the field modulation effect of the toothed-pole structure, its dimension is optimized and the performance is improved accordingly. Finally, an outer-rotor PMV machine for in-wheel motor drive is design and analyzed.

(a)

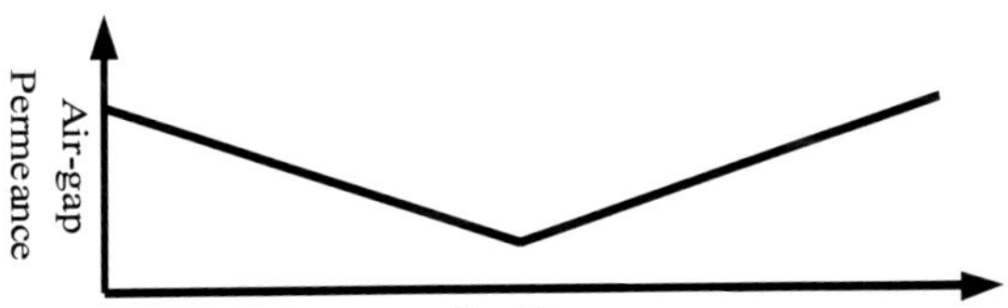

(b)

Figure 7. Vernier machine. (a) Vernier structure. (b) Air-gap permeance waveform.

Vernier Structure

The electric machine with vernier structure was proposed in 1960s [15]. The vernier structure is composed by the tooted-pole stator and moving part. Along with the motion, the air-gap permeance varies regularly with the teeth of moving part aligning or misaligning with the stator teeth as show in Figure 7. This alignment and misalignment of two sets of teeth resemble the operation of a vernier, thus the machine with this structure is called a vernier machine. A small displacement of the moving part can result in a large change of the air-gap permeance waveform which can render a high torque. Therefore, the vernier machine exhibits a high torque at very low speed which is suitable for direct-drive application.

Due to the high demand for low-speed and direct-drive application, especially with the application of high energy density PM materials, the PM vernier machines are developed in the near two decades. In terms of PM location, there are two categories, namely active rotor type (PM on the mover) and passive mover type (PM on the stator), as shown in Figure 6.

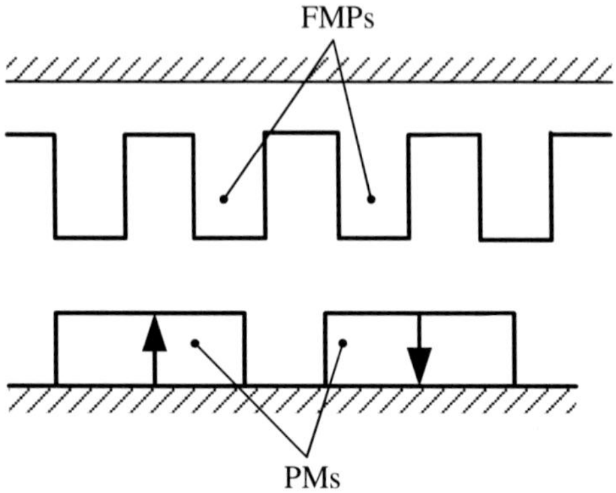

Figure 8. Toothed-pole structure.

Operating Principle

In this design, we choose the active rotor configuration. Before design, the toothed-pole structure which is also called vernier structure should be studied firstly. The concept of vernier machine lies in its toothed-pole stator configuration which resembles to the filed modulation rings in the magnetic gear, as shown in Figure 8, the toothed-pole structure. Due to the distinct difference in the permeability of ferromagnetic material and air space, the toothed-pole structure can make the different pole-pair number magnetomotive forces (MMFs) of stator and mover develop a steady force. From the point view of the magnetic circuit, the air-gap flux density can be expressed as:

$$\begin{aligned} B_{ag}(\theta,t) &= F_m \sin 2\pi P(\theta-\omega t)\cdot(\Lambda_0+\Lambda_1 \sin 2\pi Z_1\theta) \\ &= F_m\Lambda_0 \sin 2\pi P(\theta-\omega t)+\frac{F_m\Lambda_1}{2}\cos[2\pi(Z_1-P)-2\pi P\omega t]+\frac{F_m\Lambda_1}{2}\cos[2\pi(Z_1+P)-2\pi P\omega t] \end{aligned} \quad (1)$$

where F_m is the amplitude of fundamental PM MMF waveform, θ is the mechanical angle, $\Lambda(\theta,t)$ permeance function of the air-gap, Z_1 is the teeth number of the stator, and P is the pole-pair number PMs of the rotor respectively. The first item of the equation has the same pole-pair number as that of the PMs. The second and third items have much more pole-pair numbers than that of the PMs.

According to (1), the pole-pair number of MMFs and the stator teeth number should satisfy the following relationship for developing a steady torque [16]:

$$Z_2 = |Z_1 \pm P| \tag{2}$$

when ac current with an angular frequency ω is fed into the armature winding, the resultant magnetic field of armature excitation rotates at a speed of ω/P, and the rotor speed is ω/Z_2. By borrowing concept of gear ratio in magnetic gear, the gear ratio of this machine is Z_2/P. Please be noted that the above relationship is only valid for rotational machines, since its magnetic field rotating speed is P times of that of the mechanical speed.

Design Procedure

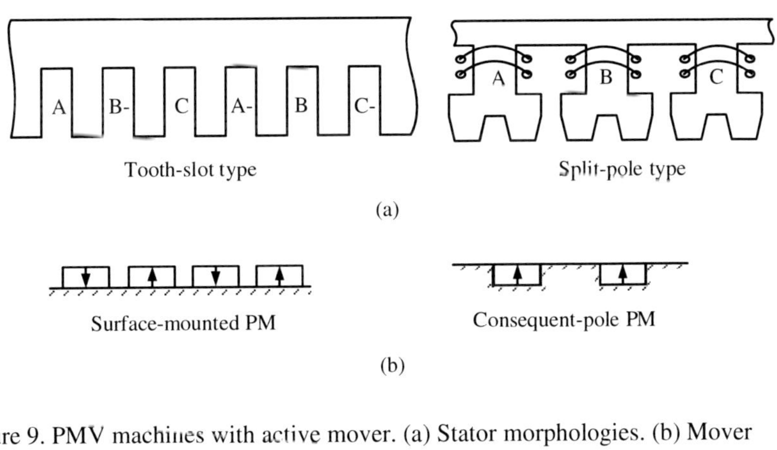

Figure 9. PMV machines with active mover. (a) Stator morphologies. (b) Mover morphologies.

Figure 9 shows the existing vernier machine topologies, only non-homopolar structures considered here. There are two main morphologies of stator: namely conventional tooth-slot type and split-pole type. The auxiliary teeth which are also termed as field-modulation poles (FMPs) in the split-pole type are split from the main teeth at the end, which can be made in a tapered or rectangular shape for field modulation. Morphologies of the mover can be divided into 2 classes according to the magnets location: namely surface-mounted and consequent-pole. In this design, the split-pole type stator and surface-mounted magnet mover are adopted. At the end of each main tooth, three FMPs were split. The advantages of the proposed generator configuration can be listed as below:

- The mover with surface-mounted magnets can be easily manufactured.
- By using FMPs for field modulation, the main slots for housing armature coils can be decreased, thus, ease the space dispute for slots and teeth.
- By using concentrated armature winding, the end-winding can be reduced, thus the copper utilization can be improved accordingly.

A. Rotor Design

The mover configuration with PMs can be classified into 3 categories, namely surface-mounted PM (SPM), surface-inset PM (SIPM), and interior PM (IPM). Each configuration has its own merits. In this design, only the SPM machine is considered which has a low-inertia and simple structure.

If the current frequency f in the armature winding is given, the rotor speed ω_m can be determined by:

$$\omega_m = \frac{60f}{P} G_r = \frac{60f}{Z_2} \tag{3}$$

where G_r is the magnetic gear ratio which equals to the ratio of PM pole-pair number and the armature MMF pole-pair number, τ is the pole-pitch of PMs on the mover. For a given operation speed and current frequency, the PM pole-pitch can be easily determined.

According to the analytical results, the fundamental component of air-gap flux density B_{ag1} can be expressed as [17]:

$$B_{ag1} = \frac{\pi}{\tau_m}\left[- a_n^{II} I_1(\frac{\pi}{\tau_m} r) + b_n^{II} K_1(\frac{\pi}{\tau_m} r)\right] \times \sin(\frac{\pi}{\tau_m} z) \tag{4}$$

B. Stator Design

According to (3), the stator tooth pitch τ_t can be calculated by the following equation with the information of the magnet pole-pitch:

$$\tau_t = \frac{2Z_2}{Z_1}\tau_m \tag{5}$$

The linear machine developed thrust force is determined by the electrical loading which limits by the cooling method. Most of the heat is caused by the copper loss expressed as:

$$p_{Cu} = 2mNI^2\rho\frac{l_{ef}}{A_c} \tag{6}$$

where m is the phase number, N is the turns of coil of each phase, I is the phase current, ρ is the copper resistivity, l_{ef} is the conductor effective length, and A_c is the area of the conductor cross section.

The heat produced by armature winding dissipates to the air space via the stator yoke. The relationship of the total copper loss with the coil temperature rise T and the heat transfer coefficient h can be expressed as [18]:

$$p_{Cu} = 2p\tau hTl_{ef} \tag{7}$$

According to (5) and (6), the machine electrical loading can be determined depends on the cooling method and its temperature limit.

C. Torque Estimation

With the information of air-gap flux density and the electrical loading in the windings, the pull-out developed thrust force can be easily estimated from the magnetic field view point:

$$T_{em} = 2PL_{ef}\int_0^{\tau_m}\frac{NI}{\Re_1}B_{ag1}\sin(2P\theta)\sin(2P\theta-\theta_e)d\theta \tag{8}$$

where $\Re_1$ and B_{ag1} is the fundamental component amplitude of the magnetic reluctance of the toothed-pole structure and the air-gap flux density respectively, L_{ef} is the stack length and θ_e is the electrical angel difference between the two magnetic field waveforms.

Therefore, given the basic design requirements, the initial specifications of the linear machine can be determined according to the above equations.

For developing a high thrust, the air-gap flux density and toothed-pole structure should be carefully designed according to (2). According to (4), the air-gap flux density depends on the PM width and thickness. A larger PM width, a higher air-gap flux density can be resulted. It can be found that the fundamental component of the air-gap flux density increases with the PM length, thus the full-pitch length is adopted.

Outer-Rotor PM Vernier Motor Drive

Figure 10 shows the machine configuration of the outer-rotor PMV motor for EV in-wheel drive. This motor adopts the split-pole structure in the outer part of the inner stator to modulate the high-speed rotating field of the armature into a low-speed rotating field so as to match that of the PMs. The armature utilizes conventional 3-phase arrangement and the armature winding connection is illustrated in Figure 11.

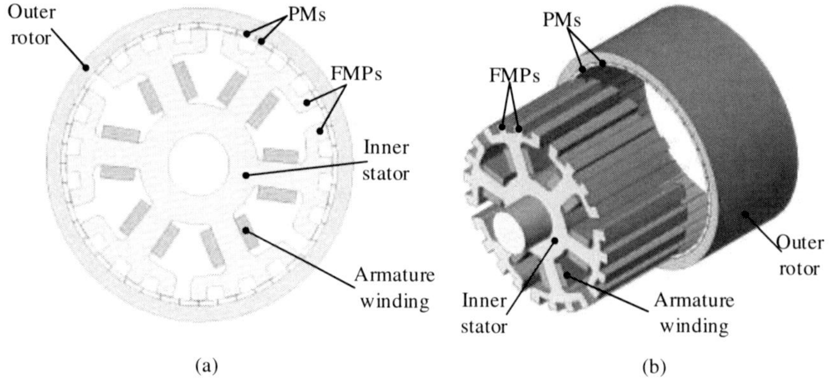

Figure 10. Configuration of outer-rotor VPM machine. (a) 2-D view. (b) 3-D exploded view.

Table 1 lists the key design data of this PMV motor. In order to assess its electromagnetic performances, the finite element method with coupled circuit simulator is adopted for calculation. Figure 12 shows the magnetic field distributions at different rotor positions. As shown in Figure 12, it can be found that as the outer rotor rotates through mechanical 8.18° in the anticlockwise direction, it corresponds to rotate through electrical 180°. It can be observed that the armature has 2 pole-pairs and the rotor has 22 pole-pairs.

Due to the FMPs, the two rotating field can interact each other to develop a steady torque.

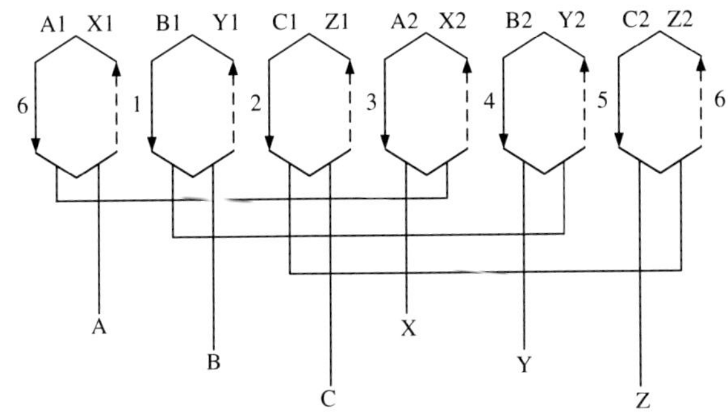

Figure 11. Armature winding connection of outer-rotor VPM machine.

Table 1. Key design data of outer-rotor VPM machine

Rated power	5 kW
Rated torque	160 Nm
Rated speed	300 rpm
No. of phases	3
No. of stator pole-pairs	2
No. of stator FMPs	24
No. of rotor PM pole-pairs	22
Rotor outside diameter	250 mm
Rotor inside diameter	217.2 mm
Stator outside diameter	216 mm
Air-gap length	0.6 mm
Axial length	120 mm
PM material	Nd-Fe-B

Figure 13 depicts the air-gap flux density. It is easily found that there are 22 pole-pairs in the air-gap which actually corresponds to three pole-pairs of the stator rotating field and 2 pole-pair fundamental harmonics, thus verifying the principle of magnetic gearing. Then, the no-load EMF waveform at the rated speed is shown in Figure 14. The no-load EMF waveform is symmetrical and sinusoidal which means that this motor is very suitable for brushless ac operation.

Furthermore, the developed torque and cogging torque waveform of the motor drive at the rated condition is numerically calculated as shown in Figure 15. The average torque of this machine can achieve 160 Nm which is sufficiently high for EV in-wheel drive. Meanwhile, the cogging torque is about 6.7% of the average torque which is very desirable for EV propulsion.

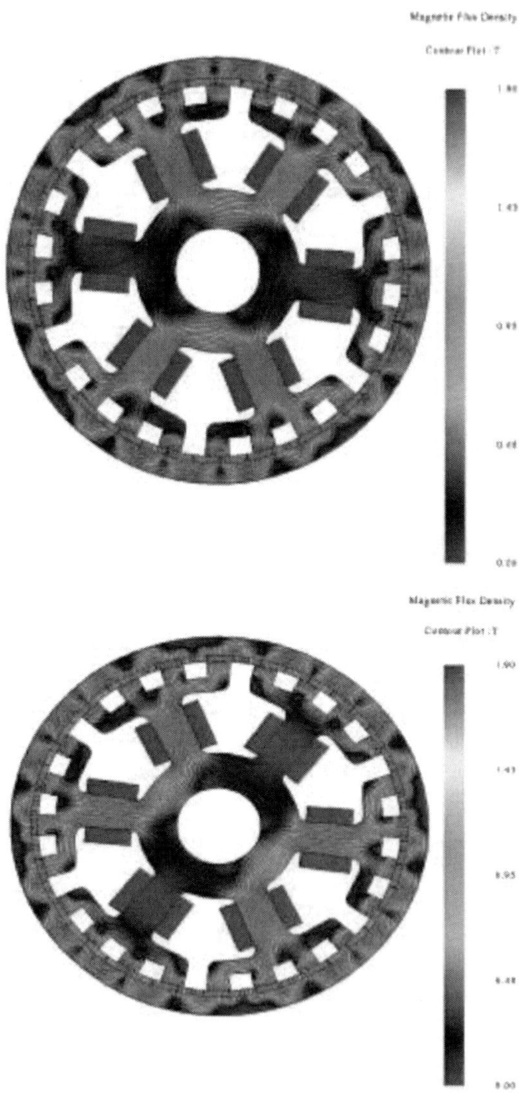

Figure 12. (Continued).

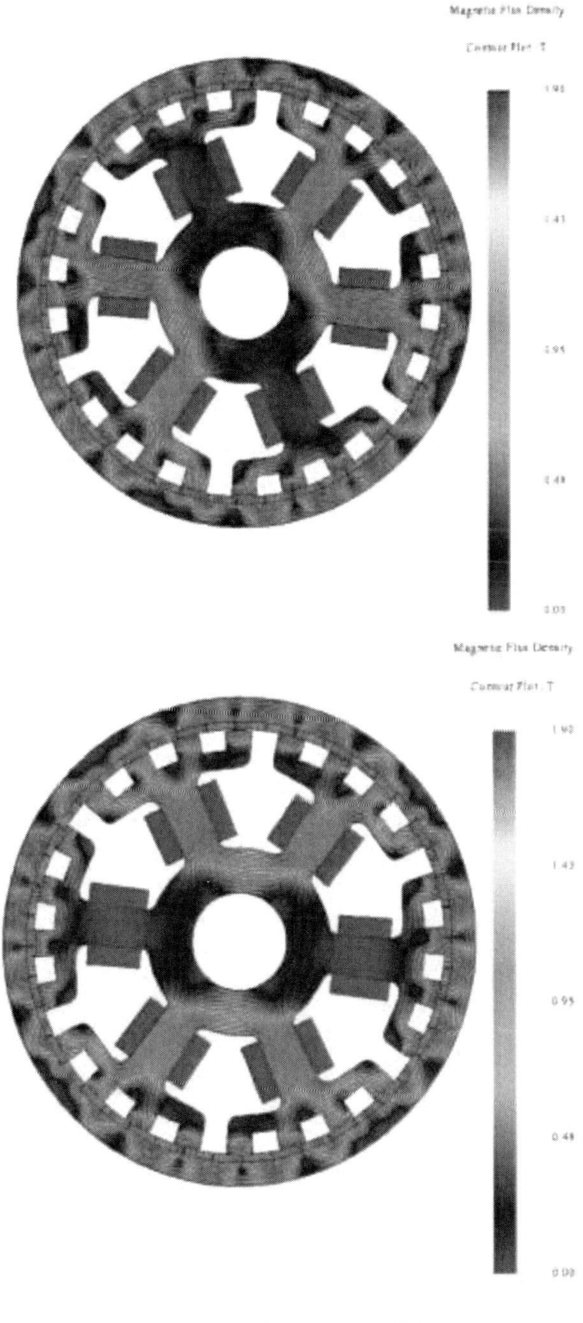

Figure 12. Magnetic field distributions of outer-rotor VPM machine at different rotor positions: (a) 0° (b) 60° (c) 120° (c) 180°.

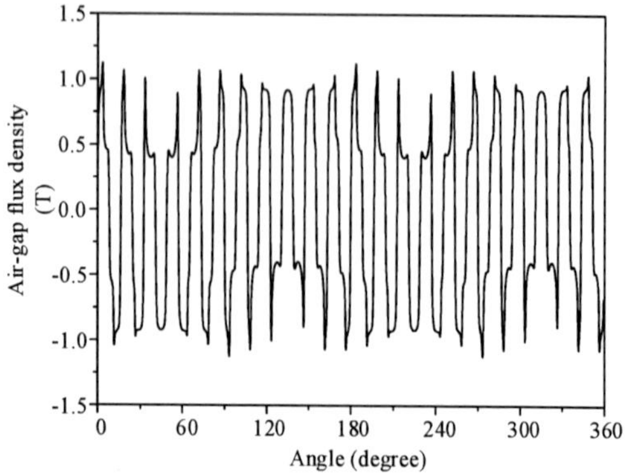

Figure 13. Air-gap flux density waveform of outer-rotor VPM machine.

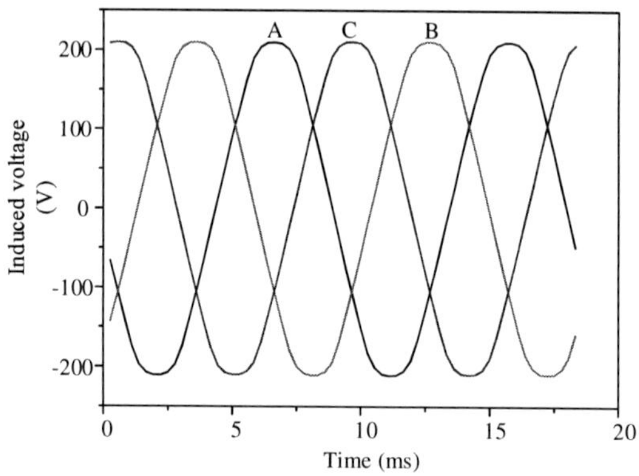

Figure 14. No-load EMF waveform at rated speed of outer-rotor VPM machine.

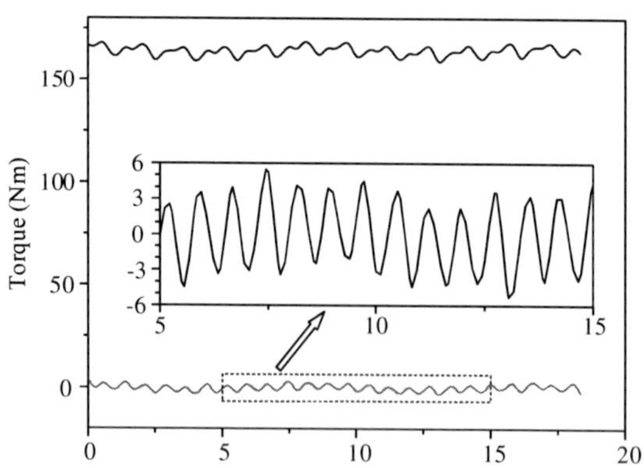

Figure 15. Torque waveform of outer-rotor VPM machine.

DUAL-ROTOR DUAL-STATOR MAGNETIC-GEARED PM MACHINE FOR SERIES-PARALLEL-TYPE HVS

In recent years, magnetic-geared (MG) permanent-magnet (PM) machines attract more and more attention due to their un-contact feature with offering gearing effect [19]-[20]. Differing from the traditional PM machines, the MG-PM machines employ electromagnetic field modulation to realize torque transmission and speed control [21]. The MG machines possess the merit of straightforward combination of the magnetic gear and the PM machine, but suffer from the drawback of three air gaps. But recently, the improved type of MG machines are presented [22]-[23]. which take the merit of involving only one air-gap, but desire a dedicated structure.

This part presents a new type of dual-rotor dual-stator (DRDS) magnetic-geared PM machine, which is proposed for series-parallel-type HVs. By connecting one rotor to the engine and coupling the other rotor to the transmission, this kind of motor can perform power splitting of the ICE output via two power converters as the planetary-gear electric continuous variable transmission in the Toyota Prius. This machine utilizes the magnetic gearing effect and can realize the non-contact torque transmission and speed variation. Therefore, high transmission efficiency, silent operation and maintenance free can be achieved.

Machine Structure

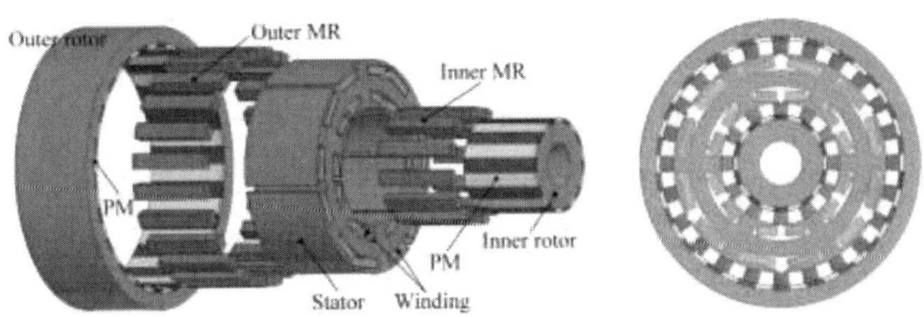

Figure 16. Proposed DRDS-MGPM machine.

Figure 16 shows the structures of the proposed DRDS-MGPM machines. It can be seen that the DRDS-MGPM type consists of two rotors, two MRs and one stator [27]. The PMs are mounted on the inside surface and the outside surface of the outer rotor and the inner rotor, respectively. The stator is sandwiched between two MRs, which also accommodates two sets of armature windings. Each MR is located between the rotor and the stator, which functions to modulate the magnetic field distribution. The outer rotor generally runs at low speeds for direct driving. So, its low-speed PM rotating field is modulated by the outer MR to the high-speed outer armature winding field. Meanwhile, the inner rotor, inner MR, and inner armature winding have the same field modulation effect. The magnetic-gearing ratios of the outer rotor and inner rotor can be designed to be the same or different, depending on the types of direct-drive applications. Finally, the inner rotor speed and outer rotor speed can be independently controlled by their respective armature windings located in the stator.

Operation Principle

Since the DRDS-MGPM machine belong to the class of MGPM machines, their pole-pair arrangements are governed by:

$$\omega_{j,k} = \frac{jp_s}{|jp_s + kN_{\mathrm{mod}}|}\omega_s, \qquad p_{j,k} = |jp_s + kN_{\mathrm{mod}}| \tag{9}$$

where $j = 1,3,5,...$, $k = 0,\pm 1,\pm 2,...$, ω_s is the rotating field speed of the stator armature winding, p_s is the pole-pair number of the stator armature winding, and N_{mod} is the number of segments of the MR for the DR-CMG machine or the number of poles of the FMP for the DR-PMV machine. In order to transmit the torque from the armature field of the stator to the rotating field of the rotor, the rotor speed ω_r and the number of PM pole-pairs in the rotor p_r should be equal to $\omega_{j,k}$ and $p_{j,k}$, respectively. When $j = 1$ and $k = -1$, the largest space harmonic component is attained for torque transmission. Thus, the corresponding air-gap flux density space harmonics and the pole-pair number can be rewritten as:

$$\omega_r = \frac{p_s}{|p_s - N_{\mathrm{mod}}|}\omega_s, \qquad p_r = |p_s - N_{\mathrm{mod}}| \tag{10}$$

In addition, the relationship between the N_{mod} and the p_s can be expressed as:

$$N_{\mathrm{mod}} = m \times i_o \times p_s \tag{11}$$

where m is the number of phases and i_o (usually, $i_o = 2,3,4,...$) is the number of MR segments or FMPs per armature winding pole per phase.

For the DRDS-MGPM machine, when the parameters of the outer MR and the outer armature winding are selected as $m = 3$, $p_s = 2$ and $N_{\mathrm{mod}} = 24$, it yields $p_r = 22$. Hence, the magnetic-gearing ratio between the outer rotor speed and the outer armature field speed is $G_r = 1/11$ which denotes that the outer rotor speed is only 1/11 of the outer armature rotating field speed in the stator. For instance, when the outer armature field speed is 2200rpm, the outer rotor speed is scaled down to 200 rpm. Meanwhile, when the parameters of the inner MR and the inner armature winding are selected as $m = 3$, $p_s = 2$ and $N_{\mathrm{mod}} = 12$, it yields $p_r = 10$. Hence, the corresponding magnetic-gearing ratio is $G_r = 1/5$. Namely, when the speed of inner armature field is 2200rpm, the inner rotor speed is scaled down to 440rpm.

Figure 17 shows the armature winding connection of the proposed DRDS-MGPM machine. Both of the two armature windings adopt the 6-slot arrangement in which the winding connections are arranged as +A, –A, +B, –B, +C, –C, +A, –A, +B, –B, +C and –C. It can be seen that this machine has larger slots in the stator to accommodate more copper conductors than the traditional PM machine, hence improving the machine power density. The key design data of the proposed machine are summarized in Table 2.

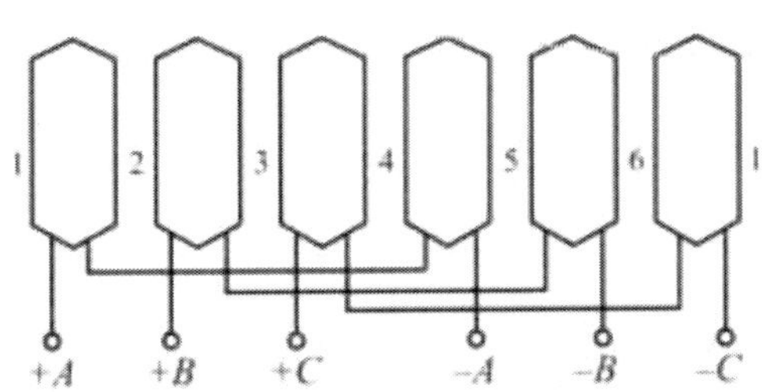

Figure 17. Winding connection diagram with 6-slot arrangement.

Table 2. Key Design Data of Proposed DRDS-MGPM Machine

Item	DRDS-MGPM machine
Outer rotor outside diameter	264.0 mm
Outer rotor inside diameter	225.2 mm
Inner rotor outside diameter	88.0 mm
Inner rotor inside diameter	40.0 mm
Stator outside diameter	200.0 mm
Stator inside diameter	111.2 mm
MR outside diameter	224.0 mm
MR inside diameter	201.2 mm
MR pitch angle	22.0 degree
FMP height	-
FMP pitch angle	-
Air-gap length	0.6 mm
Axial length	80.0 mm
Number of outer rotor PM pole-pairs	22
Number of inner rotor PM pole-pairs	10
Number of outer MR segments	24
Number of inner MR segments	12
Number of FMPs	-
Number of outer armature winding slots	6
Number of inner armature winding slots	6
Outer armature winding per slot per phase	80 turns
Inner armature winding per slot per phase	80 turns
PM material (remanence)	NdFeB (1.2 T)

Electromagnetic Field Analysis

Electromagnetic field analysis has been widely developed for electric machines. Basically, it can be categorized as analytical field calculation [24] and numerical field calculation [26]. The FEM is the one of the most popular numerical field calculation tools. The TS-FEM is developed for analyzing the proposed DRDS-MGPM machine. In general, the calculation model for the motoring mode of electric machines consists of three types of equations: the electromagnetic field equation of the machine, the circuit equation of the armature circuit, and the motion equation of the mechanical drive system. It should be noted that the calculation model involves only the first two types of equations for the generation mode.

Firstly, the electromagnetic field equations of the proposed machines are governed by:

$$\nabla \times (\nu \nabla \times A) = J \tag{12}$$

where ν is the reluctivity, J the current density, and A the magnetic vector potential. Since the end effect of the proposed machines is not significant, the two-dimensional analysis is preferred. So, both J and A have only the z-axis components. The corresponding electromagnetic field equation can be written as:

$$\Omega : \frac{\partial}{\partial x}\left(\nu \frac{\partial A}{\partial x}\right) + \frac{\partial}{\partial y}\left(\nu \frac{\partial A}{\partial y}\right) = -J - \nu\left(\frac{\partial B_{ry}}{\partial x} - \frac{\partial B_{rx}}{\partial y}\right) + \sigma \frac{\partial A}{\partial t} \tag{13}$$

$$S_1 : A = A_{z0} = 0 \tag{14}$$

where Ω is the field solution region, σ the electrical conductivity, B_{rx} the x-axis component of the PM remnanent flux density, and B_{ry} the y-axis component of the PM remnanent flux density. Along the outside periphery of the machines, the magnetic vector potential is assumed to be zero.

Secondly, the circuit equation during motoring is given by

$$u = Ri + L_e \frac{di}{dt} + e \tag{15}$$

$$e = \frac{l}{S} \iint_{\Omega_e} \frac{\partial A}{\partial t} d\Omega \tag{16}$$

where u is the applied voltage, R the resistance per phase winding, L_e the inductance of the end winding, e the electromotive force (EMF) per phase, i the phase current, l the axial length of iron core, S the conductor area of each turn of phase winding, and Ω_e the total cross-sectional area of conductors of each phase winding. On the other hand, the circuit equation during generation is governed by:

$$u = e - Ri - L_e \frac{di}{dt} \tag{17}$$

where u stands for the generated voltage. So, it can be seen that the circuit equation is similar with that for motoring, except the direction of i.

The motion equation of the machine is given by:

$$J_m \frac{d\omega}{dt} = T_e - T_L - \lambda\omega \tag{18}$$

where J_m is the moment of inertia, ω the mechanical speed, T_L the load torque, and λ the damping coefficient.

Therefore, after discretization, the three types of equations can be solved at each time step. Consequently, both steady-state and transient performances of the proposed machines can be calculated.

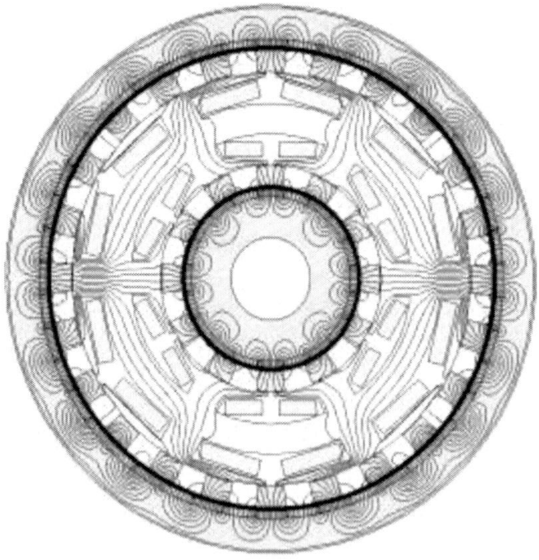

Figure 18. No-load magnetic field distribution of proposed DRDS-MGPM machine.

Figure 18 shows the magnetic field distributions of both machines under no-load condition. It can be observed that with the use of MRs, the magnetic flux of the sandwiched stator of the machine can go through the PM poles properly, hence illustrating the desired flux modulation effect.

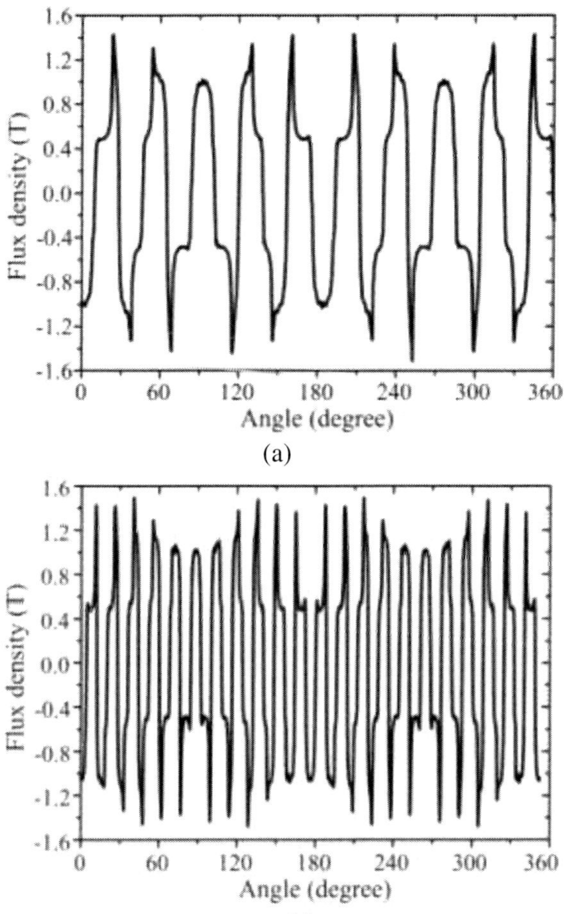

Figure 19. Air-gap flux density waveforms of proposed machine. (a) Inner air-gap. (b) Outer air-gap.

Figure 19 shows the radial air-gap flux density waveforms of the machine under no-load condition. By applying the fast Fourier transform (FFT) to these waveforms, the corresponding harmonic spectra are deduced as shown in Figure 20. It can be observed that they mainly exhibit 10 pole-pairs in the inner air-gap and 22 pole-pairs in the outer air-gap, both correspond to 2 pole-pairs of the armature winding fields in the stator. Hence, it validates the machine design described by (10) and (11).

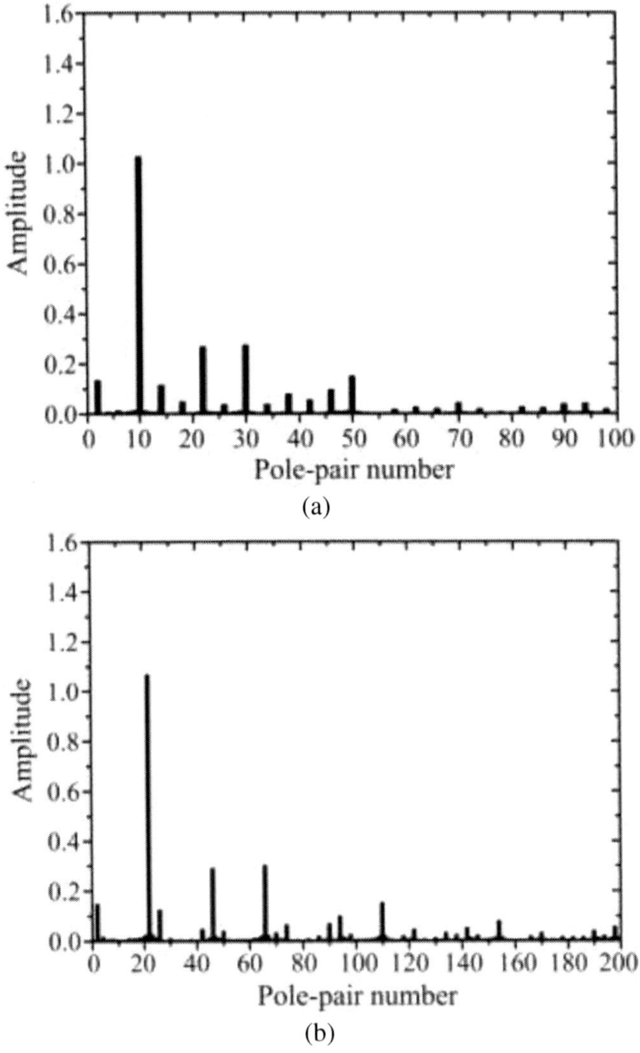

Figure 20. Air-gap flux density harmonic spectra of propose machine. (a) Inner air-gap. (b) Outer air-gap.

Performance Analysis

Firstly, the generating performances of the proposed DRDS-MGPM machine are assessed. Figure 21 shows the no-load EMF waveforms of the proposed machine under the outer rotor speed of 200 rpm. It can be observed that the machine can generate up to 54.8 V at the inner armature winding and 113.8 V at the outer armature winding. Consequently, this machine can achieve the resultant amplitude of 167.3 V.

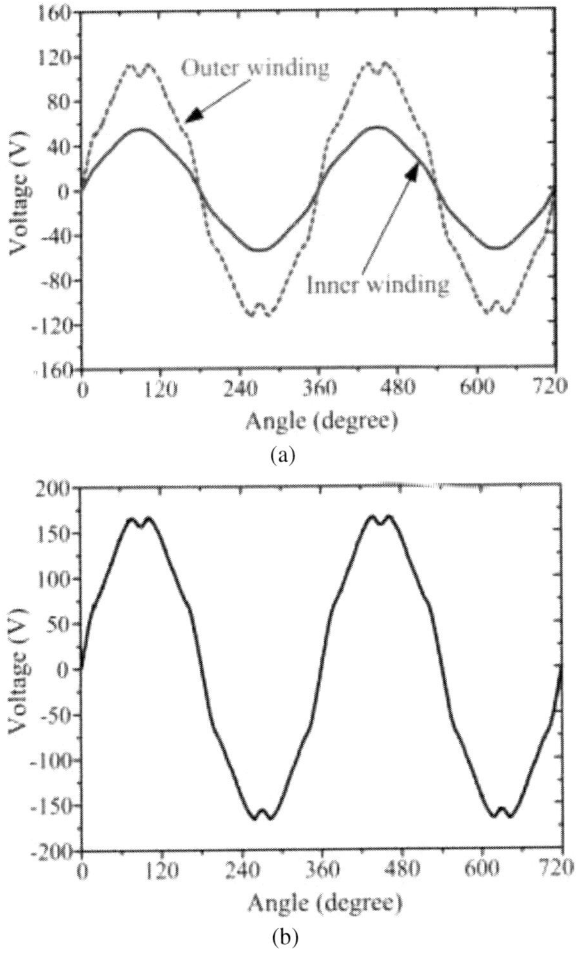

Figure 21. No-load EMF waveforms of the machine at outer rotor speed of 200rpm. (a) Separate. (b) Resultant.

Secondly, the motoring performances of the machine are assessed. Figure 22 gives the torque-angle curve of the machine. The curves are deduced by rotating the rotors step by step while keeping the phase currents at 8 A, -4 A and -4 A for both the inner and outer armature windings. It can be observed that the machine can produce the torque up to 17.6 Nm and 68.7 Nm at its inner rotor and outer rotor, respectively.

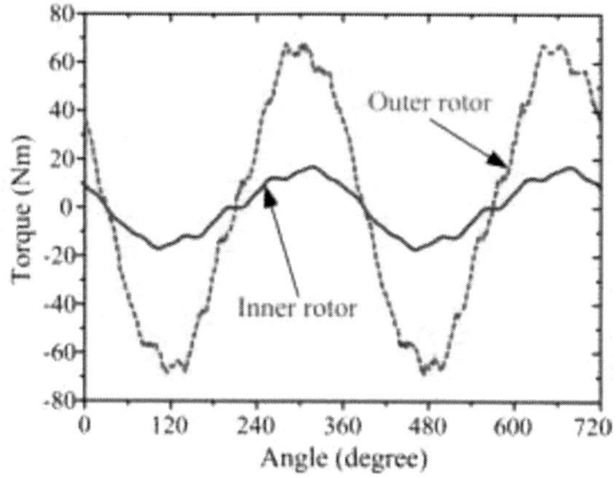

Figure 22. Torque-angle curves of proposed machine.

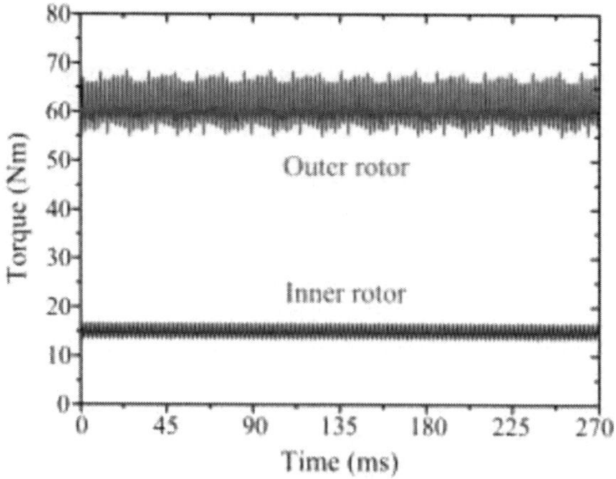

Figure 23. Output torque waveforms of proposed machine.

Figure 23 shows the output torque waveforms under the rated current of 8 A. It can be observed that the machine can produce the average torque of 16.6 Nm and 60.8 Nm at the inner rotor and outer rotor, respectively. Moreover, the cogging torques of the machine are calculated as depicted in Figure 24. It indicates that the machine exhibit the merit of small cogging torques, which is actually due to the inherit merit of multi-pole structures.

Finally, the performance comparison is summarized in Table 3. It indicates that the proposed machines can offer very high torque densities, which are up to 13884 Nm/m3 at the outer rotor of the machine. Also, the torque ripples are relatively small, which are only up to 22.0% at the outer rotor and 19.6% at the inner rotor of the machine.

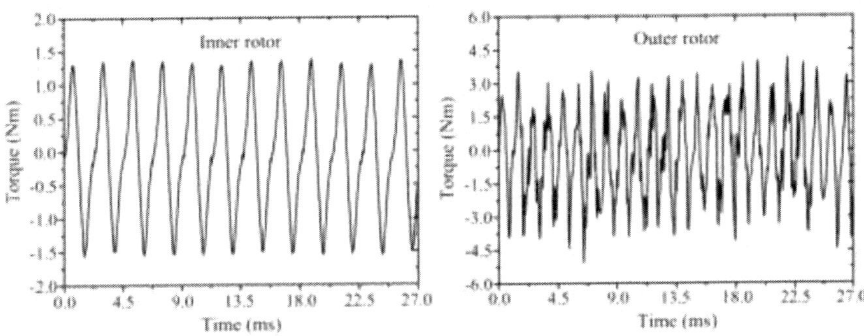

Figure 24. Cogging torque waveforms of proposed machine.

Table 3. Performances of proposed DRDS-MGPM machine

Item	DRDS-MGPM machine
Rated power	1900 W
Rated current of inner and outer windings	8 A
Rated speed of inner rotor	440 rpm
Rated speed of outer rotor	200 rpm
Rated torque of inner rotor	16.6 Nm
Rated torque of outer rotor	60.8 Nm
Maximum torque of inner rotor	17.6 Nm
Maximum torque of outer rotor	68.7 Nm
Magnetic-gearing ratio of inner rotor	1/5
Magnetic-gearing ratio of outer rotor	1/11
Machine efficiency	87.2%
Torque density of inner rotor	3790 Nm/m^3
Torque density of outer rotor	13884 Nm/m^3
Torque ripple at rated load of inner rotor	19.6%
Torque ripple at rated load of outer rotor	22.0%
Cogging torque/rated torque of inner rotor	8.3%
Cogging torque/rated torque of outer rotor	7.2%

Hybrid Stator-PM Machine for Complex-Hybrid-Type HVs

Actually, HVs needs two different kinds of power flows from electric machine for operation, namely the motoring and generating sources, which normally are performed by two electric machines, the motor and the generator. It is expected that these two machines play the corresponding roles for HV operation under different road conditions. However, if these two functions are integrated and performed by one electric machine, the total price will be significantly reduced and the room of HV will be remarkably increased.

This part presents a new type of hybrid stator-PM (HS-PM) machine, which is designed and analyzed for motor/generator dual-mode operation. It is preferred for the complex-hybrid-type HVs. With flux strengthening, the high torque can be developed for electric launch or cold cranking. With flux weakening, the constant power range of this kind of motor can be significantly extended. With the aid of flexible flux control, the output voltage is kept constant with the variable engine speed when operating as a generator.

Machine Structure

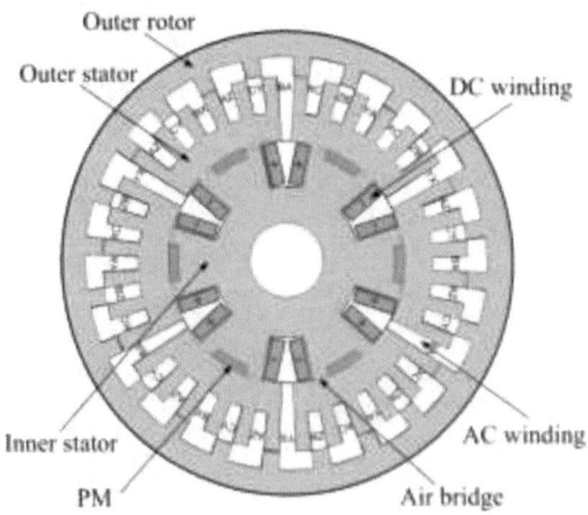

Figure 25. Proposed HS-PM machine.

Figure 25 shows the proposed HS-PM machine topologies, which indicate it having the outer rotors [28]. The proposed machine has two kind of field excitations, namely PMs and DC field windings, which settle in the inner-layer stator.

Its outer-layer stator accommodates with the AC windings, whereas its outer rotor only consists of iron steel. By controlling the DC field current, the air-gap flux density can be strengthened and weakened. Moreover, by adding the auxiliary air-bridge between the two field excitations, the regulating capability of the air-gap flux density can be further enhanced up to 9 times. The detailed configuration of this machine can be referred to [26].

Operation Principle

The proposed machine when serving for complex-hybrid-type HVs can work in the following operation modes:

Mode I: This called starting mode functions to produce the launching torque and hence bring the HV into the normal operation condition.

Mode II: This called boosting mode functions to provide the auxiliary torque to propel the HV running, such as HV climbing the hill.

Mode III: This called charging mode functions to feed the energy back to the battery, such as HV running down the hill or braking.

Mode IV: This is called the steady mode, which implies the HV working in the cruising situation. Also, in this mode, the HV is able to realize the continuously variable transmission based on the motor drive control.

Performance Analysis

The parameters and basic performances as prototype of the proposed machine are given in Table 1 and Table 2.

First, by using the time-stepping-FEM, Figure 26 shows the magnetic field distributions of the proposed in-wheel motor under flux weakening (F_{DC}=–350A-turns), no flux control (F_{DC}=0), and flux strengthening (F_{DC}=1000A-turns), respectively. It proves that the proposed machine has a good ability of flux control by tuning the bidirectional DC field current. The corresponding air-gap flux distributions are also shown in Figure 27. It can be seen that air-gap flux regulation range can be up to 9 times.

Table 1. Parameters of Prototype

Item	Proposed HS-PM machine
Number of phase	3
No. of rotor poles	24
No. of stator poles	36
No. of PM poles	6
Number of AC winding slots	36
Number of AC winding turns	46
Number of DC winding slots	6
Number of DC winding turns	150
Rotor outside & inside diameter	270.0mm &221.2mm
Stator outside & inside diameter	220.0mm & 40.0mm
Outer air-gap length	0.6mm
Stack length	240.0mm

Table 2. Performances of Prototype

Item	Proposed HS-PM machine
Rated power	6kW
Speed range	0~4000rpm
Rated torque	60Nm
Rated current	6A
Torque boost	Flux control

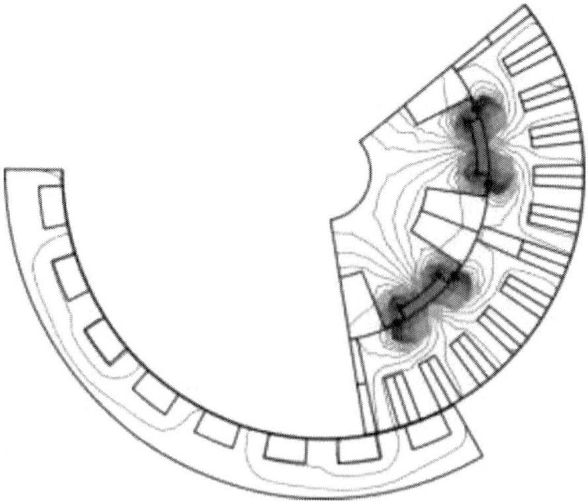

(a)

Figure 26. (Continued).

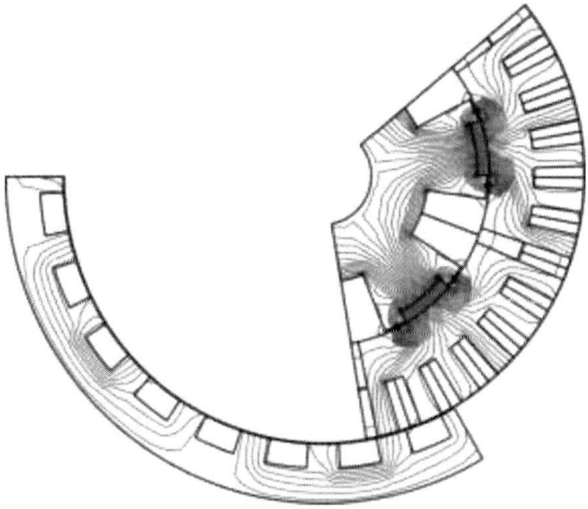

(b)

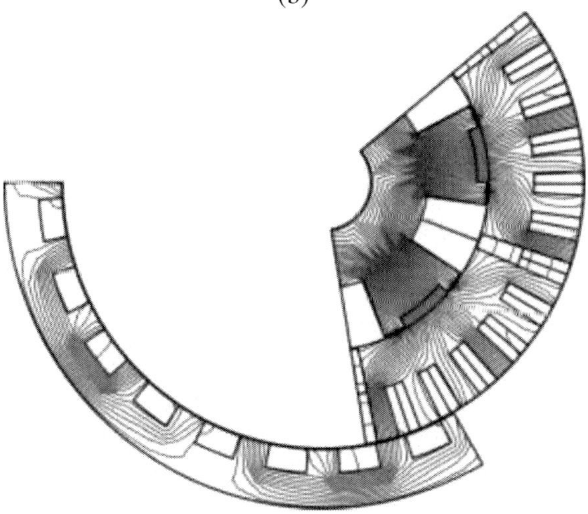

(c)

Figure 26. Magnetic field distributions under various DC field currents. (a) FDC= –350A-turns. (b) FDC=0A-turns. (c) FDC=+1000A-turns.

Second, the torque characteristics of the motor drive are assessed and analyzed. Figure 28 shows the torque-angle relationships with various field excitations. It illustrates that the torque can be effectively adjusted by flux control, hence proving this motor drive can realize a wide-speed range of constant-power operation with high efficiency, as well as offering a high torque for the HV starting.

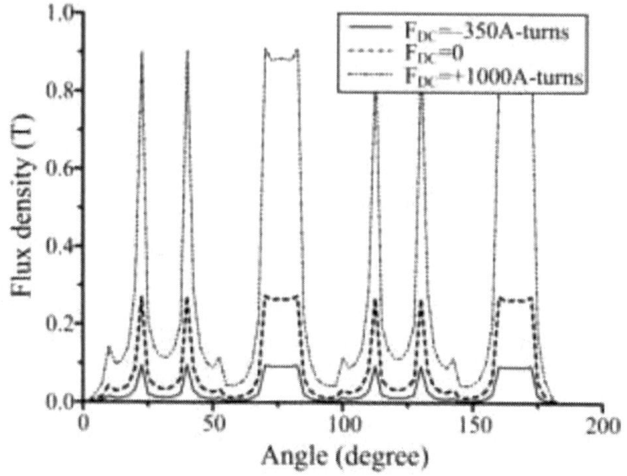

Figure 27. Air-gap flux density distributions with various field currents.

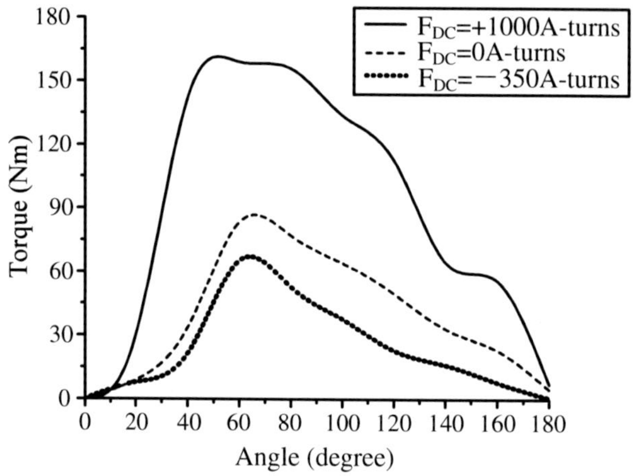

Figure 28. Torque-angle relationship with various field currents.

Third, when this machine serves in the starting mode, the transient performances with the flux strengthening under the load of 105Nm are shown in Figure 29. It can be seen that the proposed machine can offer a very high torque for the HV starting. Furthermore, the speed response shows that the proposed motor drive can provide the vehicle with an enough high speed for running, which is much challenged. Also, the starting current can be limited only about two times of the rated armature current.

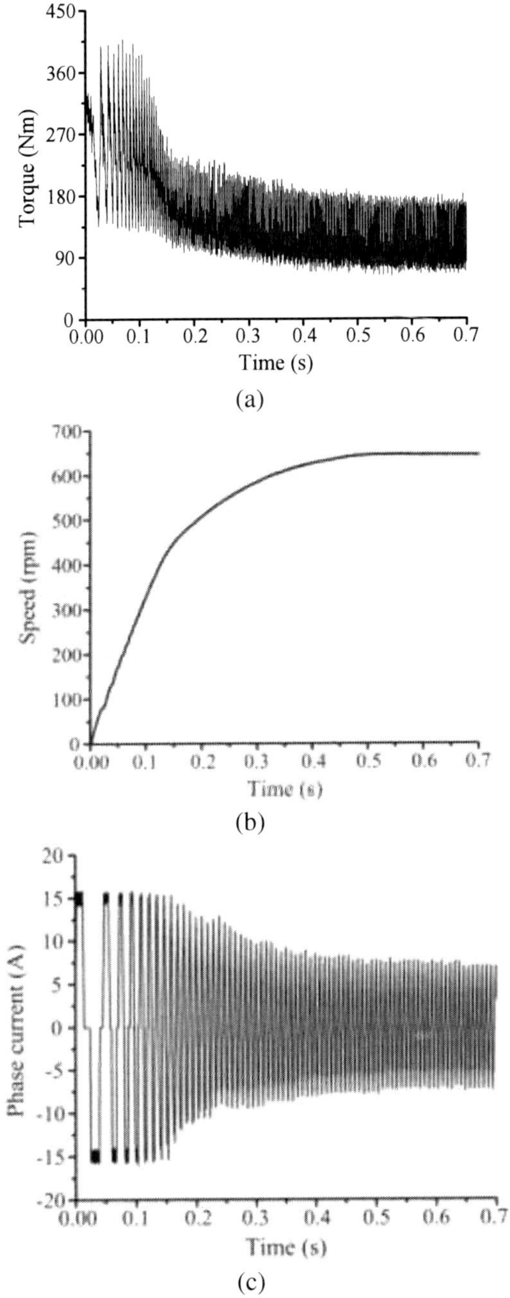

Figure 29. Transient responses with flux strengthening. (a) Torque. (b) Speed. (c) Current.

Forth, when the machine enters in the charging mode, the simulated no-load EMF waveforms without and with flux control are shown in Figure 30. It can be seen that the output voltage amplitude can be kept constant with flux strengthening at 250 rpm and flux weakening at 1000rpm. Also, the constant-voltage output characteristics entirely cover the HV working speed range of 300rpm~1000rpm, hence proving the validity of flux control.

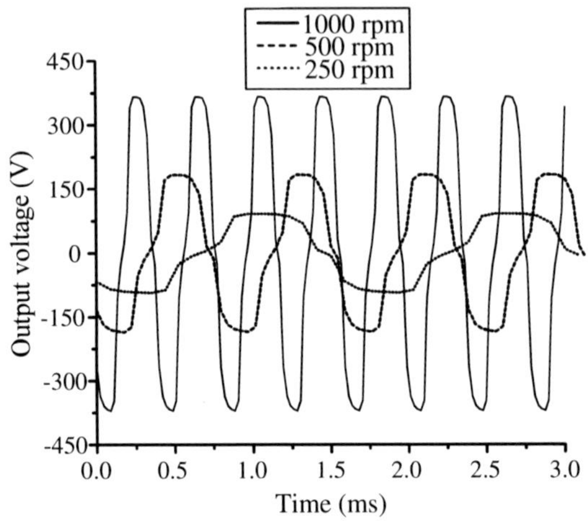

(a)

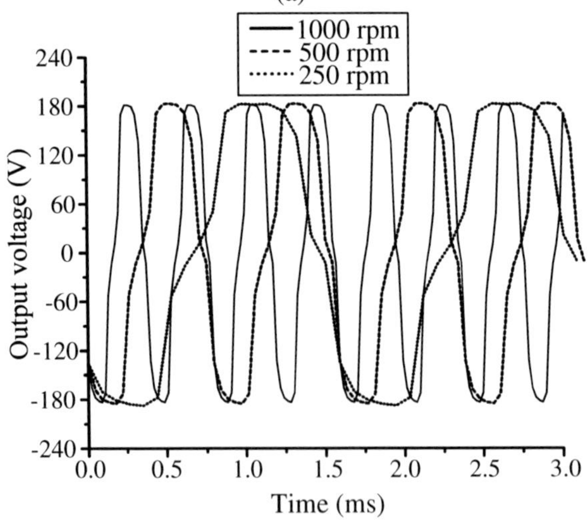

(b)

Figure 30. No-load EMF waveforms at various speeds (a) without flux control. (b) With flux control.

Conclusion

In this chapter, three emerging advanced permanent-magnet brushless machines are introduced and presented for HVs. In section I, four categories of HVs are classified, namely series hybrid, parallel hybrid, series-parallel hybrid and complex hybrid. Then, power management strategy for each kind of HVs is presented and discussed. In section II, an out-rotor PMV motor is presented and analyzed for direct-drive EV in-wheel drive which can be used in all above four kinds of HVs. In section III, a dual-rotor dual-stator magnetic-geared PM machine is presented and analyzed for series-hybrid type HVs which can be used for performing as a power-splitting unit to coordinate and control the power flow of the engine and the motor. In section IV, a hybrid stator-PM machine is presented and analyzed for complex hybrid type HVs which can be used for performing as an integrated-starter-generator to serve the multiple functions, such as engine cranking and battery charging.

References

[1] C.C. Chan and K. T. Chau, *Modern Electric Vehicle Technology*. Oxford University Press (2001).

[2] K.T. Chau and Y.S. Wong, *Energy Conversion and Management*, 43, 1953 (2002).

[3] IFSTTAR,www.inrets.fr/ur/lte/publiautresactions/fichesresultats/ficheartemis/road3/method31/All_Cycles_in_Artemis_BD_092006.xls

[4] General Motors, *Chevrolet Volt*, 2011.

[5] Honda, *Honda insight*, 2000.

[6] Toyota, *Prius Product Information*, 2014.

[7] General Motors, GM Precept, 2000.

[8] E. Spooner and L. Haydock, *IEE Proceedings of Electric Power Applications*, 150, 655 (2003).

[9] Toba and T. A. Lipo, "Novel dual-excitation permanent magnet vernier machine," *Proceeding of IEEE Industry Application Conference*, vol. 4, 1999, pp. 2539-1999.

[10] S. Niu, S.L. Ho, W.N. Fu, and L.L. Wang, *IEEE Transactions on Magnetics*, 46, 2032 (2010).

[11] S.L. Ho, S. Niu, and W.N. Fu, *IEEE Transactions on Magnetics*, 47, 3280 (2011).

[12] C. Liu, J. Zhong, and K.T. Chau, *IEEE Transactions on Magnetics*, 47, 4238 (2011).

[13] D. Yi, K. T. Chau, M. Cheng, Y. Fan, Y. Wang, W. Hua, and Z. Wang, *IEEE Transactions on Magnetics*, 47, 4219 (2011).

[14] S. Chung, J. Kim, B. Woo, D. Hong, J. Lee, and D. Koo, *IEEE Transactions on Magnetics,* 47, 4215 (2011).

[15] C.H. Lee, *IEEE Transactions on Power Apparatus and System*, 82, 343 (1963).

[16] W. Li, K.T. Chau, and J.Z. Jiang, *IEEE Transactions on Magnetics*, 47, 2624 (2011).

[17] W. Li, K.T. Chau, and C. H.T. Lee, "Optimal design and implementation of a permanent magnet linear vernier machine for direct-drive wave energy extraction," Proceeding of 38th Annual Conference of the IEEE Industrial Electronics Society, 2012, Paper No. MF-005487.

[18] N. Bianchi, S. Bolognani, D. D. Corte, and F. Tonel, *IEEE Transactions on Industry Applications*, 39, 466 (2003).

[19] K. Atallah and D. Howe, *IEEE Transactions on Magnetics*, 37, 2844 (2001).

[20] K.T. Chau, D. Zhang, J.Z. Zhang, C. Liu, and Y.J. Zhang, *IEEE Transactions on Magnetics*, 43, 2504 (2007).

[21] L. Jian, K.T. Chau and J.Z. Jiang, *IEEE Transactions on Industry Applications*, 45, 954 (2009).

[22] J. Li, K.T. Chau, J.Z. Jiang. C. Liu, and W. Li, *IEEE Transactions on Magnetics*, 46, 1475 (2010).

[23] C. Liu, J. Zhong, and K.T. Chau, *IEEE Transactions on Magnetics*, 47, 2011, 4238 (2011).

[24] W. Li and K.T. Chau, *Progress In Electromagnetics Research*, 127, 155 (2012).

[25] Y. Wang, K.T. Chau, C.C. Chan, and J.Z. Jiang, *IEEE Transactions on Magnetics*, 38, 1297 (2002).

[26] C. Liu, K.T. Chau, and J.Z. Jiang, *IEEE Transactions on Industrial Electronics*, 57, 4055 (2010).

[27] C. Liu and K.T. Chau, "Electromagnetic design and analysis of double-rotor flux-modulated permanent-magnet machines," *Progress In Electromagnetics Research*, vol. 131, pp. 81-97, 2012.

[28] C. Liu, K.T. Chau, and J.Z. Jiang, "A permanent-magnet hybrid in-wheel motor drive for electric vehicles," *Proceedings of IEEE Vehicle Power and Propulsion Convergence (VPPC2008)*, Harbin China, Paper No. H08368, pp. 1-6, Sep. 2008.

INDEX

A

B

C

D

E

F

G

P

Q

R

S

T